Developing Thinking in Geometry

Developing Thinking in Geometry

Edited by

Sue Johnston-Wilder and John Mason

The Open University in association with Paul Chapman Publishing

© The Open University 2005

First published 2005

Apart from any fair dealing for the purposes of research or private
study, or criticism or review, as permitted under the Copyright,
Designs and Patents Act, 1988, this publication may be reproduced,
stored or transmitted in any form, or by any means, only with the
prior permission in writing of the publishers, or in the case of
reprographic reproduction, in accordance with the terms of licences
issued by the Copyright Licensing Agency. Enquiries concerning
reproduction ouside those terms should be sent to the publishers.

Paul Chapman Publishing
A SAGE Publications Company
1 Oliver's Yard
55 City Road
London EC1Y 1SP

SAGE Publications Inc
2455 Teller Road
Thousand Oaks, California 91320

SAGE Publications India Pvt Ltd
B–42, Panchsheel Enclave
Post Box 4109
New Delhi 110 017

Library of Congress Control Number: 2005924502

A catalogue record for this book is available from the British Library

ISBN 1-4129-1168–0
ISBN 1-4129-1169-9 (pbk)

Typeset by Pantek Arts Ltd, Maidstone, Kent
Printed on paper from sustainable resources
Printed in Great Britain by the Cromwell Press, Trowbridge, Wiltshire

Contents

Authors

Andy Begg is from New Zealand where he has been a high school mathematics teacher and textbook author, a government curriculum officer, and a university teacher working in mathematics education. His interests include professional development and curriculum development, and the increasing emphasis on thinking in school mathematics curriculum documents impacts on both of these interests.

Geoff Faux now works freelance in mathematics curriculum development. He taught in a variety of schools before joining the Advisory service in Cumbria as mathematics advisor where he worked for 20 years. He is active in the Association of Teachers of Mathematics (ATM) and planned and convened the three-day conferences for teachers from primary to tertiary on teaching and learning reported in Mathematics Teaching (MT): Geometry (MT 129), Proof (MT 177) and Number (MT 163). He would like to remark that, of these, only 163 is a prime number. Is this a coincidence?

Sue Johnston-Wilder is a Senior Lecturer at The Open University. Originally a teacher of secondary school mathematics, she has worked with teachers and student teachers for 20 years, developing materials to promote interest in mathematics teaching and learning. She is particularly interested in teaching and learning mathematics through technology, history of mathematics and applications. She has published several books including *Teaching Secondary Mathematics with ICT* with David Pimm.

Francis Lopez-Real is an Associate Professor of Mathematics Education at the University of Hong Kong, where he has been teaching since 1994. He previously taught in schools and universities in the UK, Kenya and Brunei Darussalam, and has worked as a curriculum advisor in Cyprus and Cameroon. He has published widely in international journals and in recent years his research interests have included a focus on the impact of dynamic geometry software on the learning of geometry.

John Mason is Professor of Mathematics Education at The Open University, where he was, for some 15 years, Director of the Centre for Mathematics Education. He has had a lifelong interest in thinking mathematically himself, in working with others who want to think mathematically, and in supporting people who want to work with others to think mathematically. He has an ongoing interest in the role of mental imagery and its relation to dynamic geometry in teaching and learning mathematics, particularly geometry.

Geoff Wake is currently a lecturer in Mathematics Education at the University of Manchester with responsibility for postgraduate initial teacher training. He is committed to research and curriculum development in the teaching of mathematics through its applications. His recent work has involved the development of new qualifications that promote realistic applications of mathematics and the integration of the use of ICT in both teaching and learning. For many years he has been an enthusiastic advocate of the use of dynamic geometry software as a visualisation tool in many mathematical situations.

The writers would like to acknowledge with gratitude the support of Eric Love, Kate Mackrell, Libby Meade and Frode Rønning in the production of this book and the associated CD-ROM.

Book Series

This is one of a series of three books on developing mathematical thinking written by members of the influential Centre for Mathematics Education at The Open University in response to demand. The series is written for primary mathematics specialists, secondary and FE mathematics teachers and their support staff and others interested in their own mathematical learning and that of others.

The three books address algebraic, geometric and statistical thinking. Each book forms the core text to a corresponding 26-week, 30-point Open University course.

The titles (and authors are):

Developing Thinking in Algebra, John Mason with Alan Graham and Sue Johnston-Wilder, 2005

Developing Thinking in Geometry, edited by Sue Johnston-Wilder and John Mason, 2005

Developing Thinking in Statistics, Alan Graham, 2006

These books integrate mathematics and pedagogy. They are practical books to work through, full of tasks and pedagogic ideas, and also books to refer to when looking for something fresh to offer and engage learners. No teacher will want to be without these books, both for their own stimulation and that of their learners.

Anyone who wishes to develop an understanding and enthusiasm for mathematics, based upon firm research and effective practice, will enjoy this series and find it challenging and inspiring, both personally and professionally.

Developing Thinking in Geometry

> Shape is a vital, growing, and fascinating theme in mathematics with deep ties to classical geometry but goes far beyond it in content, meaning and method. Properly developed, the study of shape can form a central component of mathematics education, a component that draws on and contributes to not only mathematics but also the science and the arts.
>
> Marjorie Senechal, 'Shape' in *On the Shoulders of Giants*: p.139

INTRODUCTION

This book is for people with an interest in geometry, whether as a learner or as a teacher or both. It is concerned with the 'big ideas' of geometry and what it is to understand the process of thinking geometrically. By working on the contents rather than just reading the text, you will be challenged to question both your understanding of geometric thinking, and also the processes involved in working on and designing geometric tasks. The book has been structured according to a number of pedagogic principles that are exposed and discussed along the way. These are principles that teachers have found useful in preparing for and conducting lessons. If you are a teacher of mathematics, whether specialist or non-specialist, you may find that the ways of working used in this book have implications for your own teaching.

Throughout the book, tasks are offered for you to think about for yourself and, possibly after modification, to use with learners. It is vital to work on the tasks yourself so that you have experience of what the text is highlighting. We assume and expect that you will adapt the presentation of tasks to make them appropriate to specific learners' needs and experience.

Make sure that you stop and take time to work on the tasks, even if they seem too simple or too challenging at first. If they seem too simple, then you can use strategies suggested in the book to challenge yourself at your own level; if one seems out of reach, then you can try some of the strategies suggested in the book for what to do when you get stuck on a concept or a problem. In either case, these same strategies can be used with and by learners at any age.

Here then is your first task:

Task 0.1 What Is Geometry?
What does the word 'geometry' mean to you? Write a sentence or two that captures your present understanding.
Comment   Developing your ideas about what is geometry is part of what this book is about.   There is no 'right' answer to this question at this stage.

The notion of what geometry is or can be develops through the book. But it is worth stressing at the outset that we take a very definite and clear stand, namely that:

> Geometric thinking is within reach of all learners, and vital for many learners if they are to participate fully in society.

Examples of the everyday use of geometry include: moving furniture round a house, putting together flat-packed furniture, reading flood maps, using global positioning systems and interpreting perspective in art and photography.

Everyone who goes to school has already displayed the powers needed to think geometrically and make sense of the world mathematically. In order to learn geometry effectively, learners need encouragement and permission to use those powers in a supportive setting.

Furthermore, we take the view that 'geometric thinking is an absolute necessity in every branch of mathematics, and, throughout history, the geometric point of view has provided exactly the right insight for many investigations (Cuoco et al., Webref). Geometry is a part of every mathematical topic' Put another way,

> Every learner needs the opportunity to express their thinking geometrically!

STRUCTURE OF THE BOOK

Analytic geometry at school has changed its focus over the years, from traditional Euclidean geometry to the geometry of rotations and reflections known as transformation geometry. Either or both of these foci may be unfamiliar to you. Throughout this period of change, measurement has continued to be of enormous importance in practical terms. We have therefore structured the book with a block on each of Euclidean, measurement and transformation geometry.

We have also identified in the current literature and curriculum documents four key aspects of geometric thinking: invariance (for example, properties of a shape that stay the same); language and points of view; reasoning; visualising and representing. Within each block there is a chapter on each of these aspects. There are therefore 12 chapters in the first three blocks. Each of these chapters has five sections, which are based on a notional evening's work. The fifth section of each chapter addresses pedagogic issues that have arisen in the chapter.

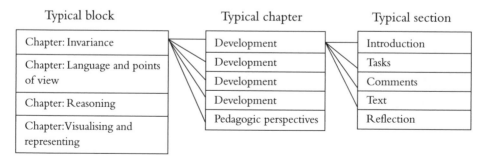

Typical block	Typical chapter	Typical section
Chapter: Invariance	Development	Introduction
Chapter: Language and points of view	Development	Tasks
	Development	Comments
Chapter: Reasoning	Development	Text
Chapter: Visualising and representing	Pedagogic perspectives	Reflection

There is a fourth, summative block of two chapters, which summarises the principles put forward in the book, the powers identified, the mathematical themes illustrated and the pedagogic constructs and strategies highlighted along the way.

ACTIVE LEARNING

Rarely is it enough just to read through materials. Learning is a process of maturation, like a fine wine or a good cheese: it takes place over time rather than just at the time of study. Full engagement comes about by actively 'doing' such things as: jotting down ideas; doing tasks and constructing your own examples; making connections; getting involved in detail; standing back to get the 'big picture'; explaining what you are doing, or trying to do, to someone else; being prepared to struggle; acknowledging feelings; and so on. We strongly recommend that you keep a notebook in which you record your work on the tasks, with a place on the side of the page for making comments and observations. You may find it helpful to record your observations of learners' responses when you try out modified versions of the tasks.

In the next task, we invite you to get started being active by thinking about A4 paper.

Task 0.2 A4 Paper

You will need a dozen sheets of A4 paper.

Lay the first sheet down on a flat surface. Lay the second sheet partly on top of the first, partly turned so that the top-left corner of the upper piece coincides with the top-right corner of the lower piece and the bottom-left corner of the upper piece coincides with the left edge of the lower piece.

Repeat this with several more pieces. What do you notice?

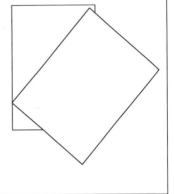

Comment

You may discover that some pieces coincide. You may be surprised at the shape you see emerging. This shape will tell you about the angles involved.

Tasks are designed to generate activity by learners in order to facilitate and stimulate interaction with the teacher. In the case of distance learning materials such as this book, interaction is with comments, with friends and colleagues, and with the learners with whom you try the tasks out.

Task 0.3 About Tasks

Look at the first two tasks again. What do you notice? In what way are they similar and in what way different?

Comment

Both tasks asked for some action from you that required you to do some thinking. The first asked you to think about your understanding, to reflect; the second invited you to engage in some mathematics. Both tasks were followed by comments that explained something of the purpose behind the task and some suggestions of other things to think about. In neither case was there a model 'solution' – what is important is your response. This is the pattern throughout the book.

How you use our 'comments' is up to you but you are advised to 'have a go' before reading them and then maybe amend or augment your reaction to the task. In this way you get a chance to engage, and maybe struggle, with the task and then set up an interaction between your thinking and the comments. Each section has a final task that invites reflection on what has gone before. No advice or suggestions are provided for these reflections.

SUMMARY

To use the book effectively, you need to engage with the tasks yourself, making observations about what you notice both in the activities and in yourself. Getting stuck for a while on a task is excellent, for it provides an opportunity to experience the creative side of mathematical thinking. If a task seems easy, then modify it so as to challenge yourself; if it seems too hard, find some way to simplify it. At the end of each chapter, think about the implications of the chapter for learning to think geometrically. You may want to try suitably modified versions of the tasks with learners, and compare their experience with your own on similar tasks.

Acknowledgements

The writers and the publishers would like to thank the following for allowing their work to be reproduced in the book.

Gilles Kuntz for permission to use CabriJava on the CD-ROM.

Olympus for permission to use DSS Player Lite on the CD-ROM.

Howard Davis for permission to use the photographs of the Gaudi Buildings in Chapter 1.

Geoff Wake for permission to use the photographs in Figures 4.0, 4.3a, 8.2b and 8.4a.

The University of Houston for permission to use Figure 4.2m and Figure 4.3c.

Northwestern University Artist, Luke Sullivan (after William Hogarth) for permission to use Figure 4.3b.

The Metropolitan Museum of Art for permission to use Figure 4.3d.

Residenz Galerie-Salzberg, for permission to use Figure 4.3i.

Multipmap.com for permission to use the map in Figure 8.2a.

Paul Bourke at the Swinburne University of Technology for permission to use Figure 8.3d.

Thespacesite.com for permission to use Figure 8.4f.

NASA for permission to use Figure 8.4g.

Nghz for permission to use Figure 8.4h.

Paul Dawkins for permission to use Figure 9.4f.

Pearson Education (Eon Harper) for permission to use the article in Chapter 10.1 'The Vulture and the Mouse' by Jon Hancock, from *Mathematics Teaching*, 185, December 2003.

In this book, there are many references to dynamic geometry software. Such software tends to display the labels of points as Roman characters instead of the more conventional Italic characters. For consistency, we have done the same.

Introduction to Block 1

There are many ideas introduced in the first block which are developed and extended in subsequent blocks.

In this first block, geometric ideas are met in the context of traditional, Euclidean geometry, which may be new to you. Euclidean geometry underpins the development of a family of modern software known as dynamic or interactive geometry software. Versions of this software are available in many schools and colleges in the UK.

In order to give you opportunity to experience this dynamic geometry software, we have created some interactive files that can be found on the accompanying CD-ROM. You will be referred to these interactive files from some of the tasks in the block.

1 Invariance

A key theme of geometry, invariance, is introduced in this chapter. In order to see, hear or feel, people need to experience both change and something to contrast with that change, namely, invariance. Consequently, invariance in the midst of change is a central theme in mathematics, and particularly in geometry.

1.1 INTRODUCTION

There are two major reasons why invariance is particularly important in teaching and learning geometry. The first is related to teaching and the second is mathematical.

Invariance and Teaching

Both the importance of experiencing variation and the pitfalls of 'limited' experiences are illustrated in the following two accounts. The accounts are written from the perspective of an adult talking with children in a primary school.

> The class had been studying the concept of parallel lines and I asked one boy if he could show me some examples of parallel lines in the classroom. He very quickly pointed out many examples to me including window and door frames, some panelling on the ceiling and the drawers of a desk. I was impressed and thought that he appeared to have a good grasp of the concept. I then happened to notice that the class had a display of flags of the world on the wall, arising from a project they had done. One of the flags was the Tanzanian national flag (as shown in Figure 1.1a).
>
>
>
> Figure 1.1a
>
> I pointed to this flag and asked if it had any parallel lines. He looked a bit puzzled and asked if I meant the edges of the flag. I agreed they were parallel but asked if he could see any others. He shook his head. I said 'What about these lines?' pointing to the yellow (diagonal) lines. He shook his head vigorously, so I asked him why not. 'Well, because they're not *straight!*' he replied forcefully.

> I sat with two girls who had been learning to name some basic geometric shapes. They had a box of flat wooden shapes and I took out a triangle, placed it on the desk between them and asked one of the girls to tell me the name of the shape. From the girl's position the triangle looked like Figure 1.1b.

Figure 1.1b

She frowned and seemed to be struggling with her answer. At last she said 'I'm not sure, but it is a triangle for *her*', pointing to her friend on the other side of the desk.

These accounts say a great deal about how children learn and how they may develop a restricted understanding of a concept from their experiences. Teachers and learners sometimes discern and discriminate differently. In the first case, it seems that the boy only recognises horizontal and vertical lines as 'straight' and possibly parallel. In the second case, the girl thinks that a triangle has to have a horizontal base in order to qualify as a triangle, even though she is aware of the possibility of looking from a different direction.

Invariance and Mathematics

A great deal of mathematics can be seen as the study of what change is permitted that leaves some relationship or property invariant (for example, the sum of the internal angles of a planar triangle is invariant as the triangle changes shape). If there is complete freedom to change (for example, investigating properties of *all* possible shapes), there may be little of any interest that can be said. When constraints are imposed on the permissible change, more interesting properties and relationships may be found. Perhaps ironically, the more constraints that are imposed, the more there is to be said about particular shapes. One example of a constraint might be to focus on quadrilaterals. Learners might investigate properties of the interior and exterior angles, and also consider whether these properties hold true for both convex and concave quadrilaterals. (See Figure 1.1c for an example of a concave quadrilateral.)

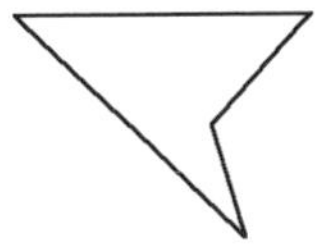

A further constraint could narrow this to studying parallelograms, kites, rectangles, cyclic quadrilaterals and so on, each in turn.

Figure 1.1c

Geometry in the World Compared with Ideal Geometry

If you look at that part of the environment described as the natural world, you are unlikely to see exact examples of geometrical shapes: squares, rectangles, parallelograms and so on are not very evident in trees, leaves and flowers, mountains or rivers. Even where shapes are recognisable, they are constructions of the human perceptual system: a full moon looks circular, but closer inspection reveals it not to be; a tree seen from a distance as having a 'shape' loses that shape as you get closer and closer; a honeycomb of hexagons becomes much less regular when viewed under a microscope. What looks like a simple geometrical shape at one magnification becomes much more complicated when you zoom in to finer detail. Even the most fundamental of geometrical objects, the straight-line segment, is hardly very evident in nature. Where does your sense of straightness come from? Perhaps it comes from gravity imposing verticality on your experience, and from your sense of 'not deviating' when walking.

The constructed environment abounds in geometrical shapes, especially rectangles and triangles. Indeed, you live in a very 'rectangular' world (as opposed, for example, to a hexagonal or circular world). When architectural design follows the more fluid and organic shapes of nature, you may be taken aback by its strangeness. Look at the photographs of Antonio Gaudi's Casa Batlló and Casa Mila in Barcelona (Figures 1.1d and 1.1e overleaf) and notice your reaction. (See also Gaudi webref).

Figure 1.1d

The world of geometry as handed down from the Ancient Greeks, and known as Euclidean geometry, is a world that is entered through use of the human power to imagine. It is a conceptual world, filled with 'ideal' shapes. Euclidean geometry is not concerned with measurement in the material world but, rather, with exact relationships that, by their very nature, are conceptual. Thus, when you say that the lengths of the four sides of a rhombus are equal, you mean exactly equal. For any particular figure drawn on paper say, and intended to represent a rhombus, you can only ensure the equality of its sides to a degree of accuracy dictated by the tools of construction.

Figure 1.1e

Seeing

Metaphorically the expression 'I see what you mean' is used in English to mean 'I understand'. This 'frozen metaphor' is so deeply embedded in language that it usually escapes notice, but it illustrates how much visual perception is considered to be an integral part of learning and comprehension. Some people respond more fully to 'I hear what you are saying', for their preferred mode of comprehension is verbal and aural. The study of visual perception has long been a fascinating aspect of psychology and you may have come across Kanizsa figures like the one in Figure 1.1f (named after Gaetano Kanizsa who first studied them).

Figure 1.1f

The 'triangle' that you see is a construction of your perceptual system. It is not on the page! Such illusory figures illustrate that 'seeing' is not a passive activity but involves active construction on the part of the viewer. In this case, even very young infants 'see' a triangle. A central feature of geometry is learning to 'see', that is, to discern, geometrical objects and relationships, and to become aware of relationships as properties that objects may or may not satisfy.

Reflection 1.1

Try telling someone else what you understand by 'invariance in the midst of change'. Find some examples from the teaching that you are currently doing or planning.

1.2 DISCERNING SHAPES WITHIN SHAPES

This section is about discerning different shapes, and relationships between shapes, in figures. You will experience mentally shifting, rotating and flipping objects in order to relate them to some other object. While you are doing so, keep looking out for examples of invariance amidst change.

Discerning Triangles

The first task in this section uses a figure often found in texts about the theorem attributed to Pythagoras.

Task 1.2.1 Discerning Triangles

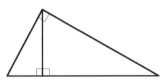

Figure 1.2a

Look carefully at Figure 1.2a. What do you have to do in order to discern three triangles?

Comment
At first the outer triangle may stand out, or you may be aware of the two inner triangles. To attend to the outer triangle you have to attend to the overall shape and ignore the inner segment; to attend to the inner triangles, you have to stress the inner vertical line segment and ignore the overall shape. It is vital to be aware of all three triangles (and relationships between them) in order fully to appreciate the diagram.

It is not always easy to discern sub-shapes or embedded shapes in more complex figures. For example, many of the most familiar logos in the environment are based on simple geometric designs, such as the HSBC logo shown in Figure 1.2b.

Task 1.2.2 HSBC Logo

How many different pentagons can you find in Figure 1.2b?
How many different polygons can you find in the figure?

Figure 1.2b

Comment
The core of the task is to pay attention to how you discerned different pentagons.
  Did you find a concave pentagon?
  Did you find yourself asking 'what does it mean to be different'? For example, does colour (shading) matter? What is important is to become aware of this as an issue, and to make your own decision. Different decisions will produce different answers.
  How did you go about recording your results? A natural way would be to highlight each identified shape by shading or thickening the edges. If you decided that different orientations or positions of *congruent* shapes are significant then, for example, you would immediately identify six isosceles right-angled triangles. If you consider these to be the same because of their *congruence*, then there is only one such triangle. (The meaning of the term *congruent* here describes shapes which, if cut out, can be made to fit exactly on top of each other.)

An issue that often arises in geometry is how to record your observations in a form that will convince others. For example, you could shade in one trapezium as shown in Figure 1.2c. The shading is there to suggest stressing the shaded figure and ignoring the other line–segments.

Figure 1.2c

The number 4 indicates that there are four such congruent trapezia (or trapeziums) in different orientations. Of course, you need to be clear in your own mind what these orientations are. If you were presenting your record of results to someone else, such a record would mean that you expect the reader to do some work on the image too.

Another aspect raised by 'different' shapes in the problem is the question of *similarity*. Isosceles right-angled triangles occur in two different sizes in the logo, but they are the same *shape*. So your definition of 'different' might encompass non–similar as well as non–congruent.

Three important issues emerge: discerning, labelling and counting depend on deciding what counts as different; making a choice for yourself is usually motivating; most problem–solving situations, particularly real–life problems, involve deciding what matters and what does not, and so making choices.

Apart from the question of recording your results, how did you go about *finding* different shapes? You may have started out fairly haphazardly, picking out shapes that jumped out at you. Each shape is identified by stressing or discerning some features and ignoring others. You probably found yourself needing to be more systematic in your search, leading to some kind of classification. Such processes (that is, classifying and being systematic) are common features of problem–solving and mathematical thinking. In this particular example, you could have decided to organise your search by systematically increasing the number of sides of the polygons, starting with triangles. An alternative approach could be to *remove* parts of the diagram systematically. That is, you might notice at the outset that the whole diagram is composed of six small triangles and so start with the whole hexagon and ask how many different shapes can be made by removing one triangle at a time, then remove two triangles, and so on. If you were to take this approach, you would quickly encounter another decision regarding definitions. For example, removing two triangles produces the shape shown in Figure 1.2d, which could be perceived as having six vertices and six sides, because two sides cross in the middle but do not form a vertex there. Such a shape might be called 'self-crossing'. A decision is required as to whether it is counted.

Figure 1.2d

In discussing this task, two important elements of learning can be identified. The first relates to the *content* of the task, that is, the geometrical objects and relationships that are specific to the task. In this case, this would include particular shapes like the trapezium, more general ideas like convex and concave polygons, and relationships like similarity and congruence. The second element relates to the *processes* involved in solving the problem. These may apply to a wide range of problems, not just this specific problem or just geometric problems. George Polya, a famous Polish-American mathematician, wrote several books full of excellent advice about teaching and learning mathematical processes. For example, he suggests inviting yourself to 'check the result' and 'derive the result differently'.

There are many problems in mathematics where checking the result is a relatively simple matter, but counting is not so easy: how can you be sure that you have not missed any shapes? In more complex problems of this type, can you be sure that you have not double counted some of the shapes? This is one of the reasons that being systematic is valuable. Your system needs to be robust enough and logical enough to give you confidence that you have solved the problem completely. Another approach to counting is to try to find a different way of counting to see if you get the same answers. The next task offers an opportunity to work on counting in a complex diagram.

Task 1.2.3 Seven Circles

How many different pentagons can be made by joining intersection points of the circles in Figure 1.2e? Here 'different' means not congruent and not similar.

For each pentagon, describe how to work out the angles without measuring.

Consider other polygons as well.

On the CD-ROM, there is a worksheet (**Worksheet for seven circles**) on which the diagram is replicated, which you can print out and use to record your work.

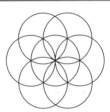

Figure 1.2e

Comment

This task has similarities with the previous one but, instead of the diagram consisting of polygons, the task effectively provides you with a set of points (determined by the seven circles). You have the freedom to select the points you want to work with.

Equilateral triangles feature quite significantly (the circles force equal lengths and hence equal angles), so you may wonder how many different sizes of equilateral triangle can be found. This focuses attention on similarity, suggesting looking for other triangles that are similar, but not congruent, to each other.

How did you go about determining angles? Your starting point might be the 60° angles that are formed at the centre of the diagram. Using this fact, together with all the equal radii in the diagram (thus producing many equilateral and isosceles triangles) might enable you to find the angles for most of the polygons that can be drawn. This is an example of how reasoning involves using perceived properties to deduce further properties of identified objects.

If you were to extend your investigation beyond pentagons to consider any polygons, you would need to make decisions about what to focus on. This raises more geometrical issues. For example, triangles with corresponding angles equal must be similar, but for rectangles, which can vary from extremely long and thin to square, having corresponding angles equal is not sufficient to ensure similarity. How then might you convince someone else whether the two rectangles shown in Figure 1.2f are similar or not?

Are you even sure that the two figures are rectangles? Are their edges actually straight? Perhaps some angle calculations would make it convincing.

Figure 1.2f

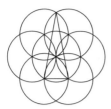

Figure 1.2g

At first sight, it seems that it should be possible to find all the angles in all the polygons, and, moreover, that these angles will be simple fractions or multiples of 60°. However, Figure 1.2g shows a triangle whose angles cannot be found without use of trigonometry. The fact that rectangles can be drawn in the diagram leads naturally to the question 'Is there a square?' As usual, it depends on what you decide to permit: are further construction lines allowed? These elaborations serve to illustrate how rich this figure is and how it could be used fruitfully with students having a wide range of mathematical knowledge and background.

The next task is similarly rich in possibilities.

Task 1.2.4a The Nine-pin Geoboard

A nine-pin *geoboard* has nine pins laid out as shown in Figure 1.2h. Coloured elastic bands can be used for outlining shapes.

How many different triangles, quadrilaterals and so on can you find? How many different angles can you find?

(Instead of using an actual geoboard you could join points on copies of the diagram shown in Figure 1.2h.)

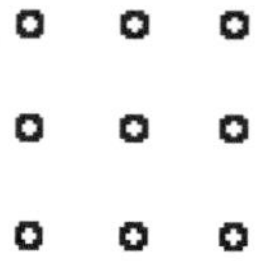

Figure 1.2h

Comment

Were you systematic? Did you catch yourself discerning some detail, seeking relationships or reasoning on the basis of properties?

If you were thinking geometrically, you did not bother about the fact that the elastics and the pins have thickness, so that angles are very hard to measure physically. You concentrated instead on the ideal shapes indicated by the physical shapes that you imagined.

These tasks also provide experience of the pedagogic difference between analysing something already given, and constructing your own objects within specified constraints.

Relating Areas and Angles

The nine-pin geoboard is an excellent resource for working on relationships between areas of shapes.

Task 1.2.4b The Nine-pin Geoboard (Areas)

Find relationships between the areas of the polygons you found in task 1.2.4a, and particularly in terms of the area of the outer square determined by the nine pins.

Comment

The word 'area' may summon up formulae, but it is perfectly possible to see each of the polygons as made of pieces that are recognisable fractions of the outer square.

The outer square forms one 'unit' of area, against which all the others can be compared. You may prefer to use the area of the smallest square formed by the pins as your unit of area. It would also be possible to use the area of a triangle made up of three adjacent pins forming a right-angled triangle.

Seeking relationships between areas of shapes without actually calculating areas requires you to discern elements and relate them by mentally juxtaposing or partitioning them. For example, in the first diagram in Figure 1.2i, there is a relationship between the areas of the shaded triangle and the enclosing rectangle. Mental partitioning reveals the area of the shaded triangle in the second figure to be half that in the first. The third and fourth triangles can be treated similarly. The important point to stress here is that the relationship between the area of a triangle and a rectangle having the same (or equal) base and height as the triangle is an *absolute* relationship. That is, it is independent of any specific measure of area that might be used.

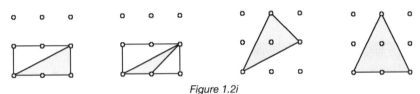

Figure 1.2i

Many learners claim that the fourth triangle shown in Figure 1.2i is an equilateral triangle and therefore that its angles are 60°. It does *look* like an equilateral triangle, but by attending to edge lengths, you can see that the base is shorter than the other two sides. This raises a question about its angles.

Task 1.2.4c The Nine-pin Geoboard (Angles)

Find one or more ways of relating different angles in the triangles without actually measuring them.

Comment
Familiar angles of 90° (one-quarter of a revolution), and 45° (one-eighth of a revolution) are easy to spot. Mathematicians use a variety of measures of angle including fractions of a revolution, but a natural geoboard measure is to describe angles in terms of vertical and horizontal displacements as described below.

The first two angles in Figure 1.2j could be described as [1,2] and [2,1].

Figure 1.2j

Notice that you would need to specify which displacement is in the first position, and then stick to it. This may well remind you of other notations in mathematics, for example, co-ordinates of points in the plane, or vectors in the plane. If you are familiar with trigonometry, you may also realise that the description of the first angle is another way of recording 'the angle whose tangent is $\frac{1}{2}$'.

Using this system, the angle 45° can be recorded as [2,2] or [1,1], which implies that [2,2] = [1,1]. How would 90° be represented in the system? Or 0°? What about the other two angles in the Figure 1.2j? The fourth angle could be described as twice [1,2]. This might be written as 2[1,2] but notice that this is *not* the same as [2,4]. Similarly, the third angle could be described as [2,1] − [1,2] but there is no easy way to simplify this expression.

In this particular case, the value of such an investigation is not in the 'results' obtained (you might well conclude that the system is too clumsy to be of practical use) but rather in the experience of mathematical invention.

By engaging in the creative activity of trying to find a good way to compare angles, learners can experience the complexity of the problem, the usefulness in different contexts of thinking in different ways, and the fact that the angle measures commonly in use are a convention.

Reflection 1.2

This section has introduced the ideas of discerning features and seeking relationships between figures in relation to shape, area and angle. It has provided experience of mentally shifting, rotating and flipping objects in order to relate them to some other object. How do you see these drawing upon the theme of invariance in the midst of change?

1.3 BUILDING UP AND SHIFTING SHAPES

Although some of the tasks in the previous section invited you to 'construct' your own shapes, nevertheless you have still been severely constrained by the configurations that you have worked with. The tasks in this section give you much more freedom. In the course of the section you will encounter transformations of shapes, each of which has associated invariance.

Task 1.3.1 Combining Two Congruent Triangles

Cut out two identical triangles and place them together along equal edges. How many different shapes can be made? Describe properties of the shapes formed.

Comment
You can do this task in the practical way suggested, by actually cutting out the shapes from card. Many people find the handling of real objects a satisfying and insightful way to investigate geometrical ideas. If you wish, you can open the interactive file '**1a Two Triangles**' on the CD-ROM and manipulate the triangles on the computer screen using a mouse. Start with scalene triangles and then consider some special cases such as equilateral, isosceles, right-angled and obtuse-angled triangles and combinations of these (for example, isosceles right-angled).

The task does not mention whether you are permitted to 'flip' one of the triangles. Of course, as you know by now, this is entirely up to you. If you do not allow flipping you limit the possibilities. For example, flipping a scalene triangle doubles the number of shapes you can make. However, you will also have found that only two 'types' of quadrilateral were formed: parallelograms and kites. What is most interesting is to work on explaining why quadrilaterals are always formed. Constraining yourself to special triangles produces special quadrilaterals.

Instead of using two congruent triangles, you could use two triangles that have just two matching pairs of sides or two triangles that have just one matching pair. You might forecast, from your previous activity, how many different shapes can now be formed. From this experience, you may feel prepared to forecast the number of different shapes formed if you place together two congruent quadrilaterals, and perhaps even to generalise for any pair of congruent polygons. Moreover, because there are many 'special' quadrilaterals (for example, parallelograms, rectangles, kites, rhombuses, squares and so on), the exploration is likely to be richer than with triangles.

The suggestion about trying to combine triangles with just one matching pair of sides has an interesting connection with another problem. When studying perimeters and areas of shapes, learners often wonder whether having equal perimeters implies equal areas and vice versa. Many children believe this is true before they investigate. Notice that this is essentially a question about whether something is invariant when something else is changed. However, the following question from a pupil may be one that you have not thought about:

> If two shapes have the same perimeter *and* the same area, does this mean they must be congruent?

The question refers to shapes in general, so probably a useful strategy is to focus on particular shapes to begin with, for example, triangles or quadrilaterals. Try out some particular examples in each case, although sometimes it is not easy to check. An important strategy when testing a conjecture is to see if you can find a counter-example. In this case, the previous task immediately provides an answer. When two triangles with just one equal side are put together there are two possibilities, illustrated in Figure 1.3a.

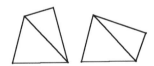

Figure 1.3a

Both quadrilaterals in Figure 1.3a have the same area and perimeter but they are not congruent. You might like to try giving this problem to learners and see how they approach it.

One of the most famous activities that involves building shapes from given pieces is the ancient seven-piece Chinese Tangram puzzle (see Figure 1.3b). The 'puzzle' is usually presented in the form of a dissected square and the pieces then have to be reassembled to form a variety of shapes (both realistic and geometric) for which the silhouettes are given (see Tangram puzzles webref). The geometric thinking involved has to do with the invariance of area under moving of the pieces, as well as trying to imagine a silhouette shape as decomposed into the puzzle pieces.

Figure 1.3b

Moving and combining shapes is preliminary to moving parts of shapes to create new figures or to consider relationships between parts of a figure.

Task 1.3.2 Seeing As

In the two stars shown in Figure 1.3c, try to see each star as created in as many different ways as possible.

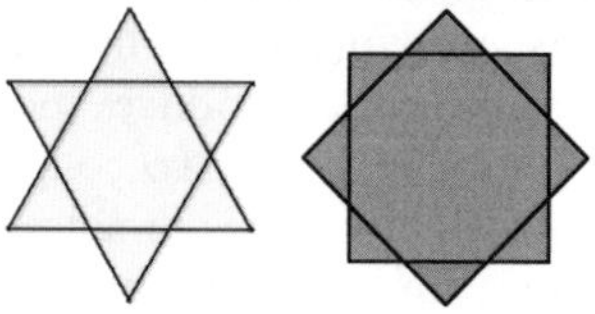

Figure 1.3c

Comment

Max Black (1993) suggests this type of task as a way to prompt learners to try to see things differently before plunging into the first way of seeing that comes to mind. Different ways of seeing often afford access to different ways of thinking. People have seen the first star as, for example, a small hexagon with triangles on its edges, two overlapping equilateral triangles or three overlapping rhombuses.

You could also work at finding all the different shapes embedded in the figures. The first star produces many of the same shapes that you found in the seven-circles task.

Task 1.3.3 Shifting Pieces

Think of the stars above as two overlapping shapes – triangles (or squares). What changes and what stays the same when you:

(i)   shift one of the triangles (squares) in a vertical direction?

(ii) shift one of the triangles (squares) in a horizontal direction?

(iii) rotate one of the triangles (squares) about its centre?

Comment

Once again, you have a choice about how to work on the task: you can cut out two shapes, drawing one shape on paper and the other on a transparent sheet to be overlaid on the first, or you can use the interactive file '**1b Star**'.

In addition to the separate shifts and rotations described, you could look at combinations of these movements, but this becomes more complex to describe and analyse.

Apart from 180°, there are only two different angles in the first star diagram (60° and 120°) and only three in the second (90°, 45° and 135°). After experimenting with horizontal and vertical shifts, you will notice that all angles remain invariant.

Imagine a straight line being translated while another straight line that crosses it remains invariant. Then the angles between the fixed line and the translated line remain invariant. The diagram on the left in Figure 1.3d shows the situation for a horizontal shift of a straight line. This means that the only changes that occur with the stars are in the lengths of sides in the overlap region.

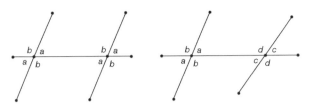

Figure 1.3d

In the case of rotating one of the shapes, rotating a line segment usually produces a change in direction. So, focusing on the intersection of two segments and rotating one of these segments must produce a change in the angles of intersection as shown in the diagram on the right of Figure 1.3d. What remains invariant under rotation? The angles and lengths of the rotated figure do not change, just the angles and distances between rotating and fixed segments. So, thinking in terms of triangles or squares, these figures remain invariant as shapes, while the figures formed from their intersections change as one of them rotates while the other stays fixed.

There are further invariances when one of the shapes is rotated about its centre. If the vertices of the inner figure (hexagon or octagon) are joined to the centre, it is composed of congruent triangles, which might remind you of Task 1.3.1. The diagrams in Figure 1.3e each illustrate congruent triangles placed together in two different ways.

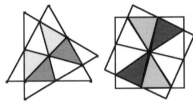

Figure 1.3e

Reflection 1.3

What remains invariant about a shape when it is translated?

What remains invariant about a shape when it is rotated about some point?

What remains invariant about the intersections of figures when one shape is translated (or rotated) and another remains fixed?

1.4 CONSTRAINING RECTANGLES

In section 1.1, it was remarked that the human-made environment is predominantly 'rectangular'. This section concentrates on rectangles (and on squares as a special case) but with a focus on the behaviour of these shapes when they are subjected to certain constraints.

Task 1.4.1a Square and Rectangle

In Figure 1.4a, ABCD is a square and E is a point on AB such that AE > EB. In one figure E is located to make a square, and in the other a rectangle with the line BD a line of symmetry.

Which inner shape has the greater area, the square or the rectangle?

How many different ways of reasoning can you find? Which do you find most compelling?

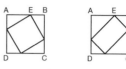

Figure 1.4a

Comments

It is important here that you experience a choice of approaches.

Since no lengths are shown in the diagrams, you do not know precisely where the point E lies, only that it is somewhere beyond the mid-point from A to B. It is possible to adopt a strategy of 'drawing and measuring', however, this would only give you an answer for some specific positions of E.

The generality of the problem posed suggests that a general conclusion might be possible. You could try extreme cases (E at the mid-point, or at B). You could cut out a square and fold, first along the sides of the inner square and look at areas, and then along the sides of the inner rectangle, and look at areas. You could also imagine folding and draw the result. If you would like to, try the interactive file '**1c Square and Rectangle**'.

Whichever approach you used, you would need to think about *how* the inner square and rectangle can be drawn inside the square, and how they are related. This might lead you to look at the other shapes within the diagrams (for example, the triangles around the square and the rectangle) and to think about relations between them. Observation and symmetry would suggest that the four triangles around the inner square are all congruent, while around the rectangle there are two pairs of congruent isosceles triangles. If you imagine folding the outside triangles inwards, you might see in your mind a diagram like one of those shown in Figure 1.4b.

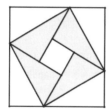 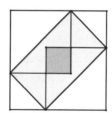

Figure 1.4b

In the first case, folding the four triangles inwards leaves a space in the middle from which you can conclude that the area of the four triangles together must be less than half of the large square. Thus, the area of the inner square must be more than half of the area of the large square.

In the second case, the triangles overlap. Thus, the area of the rectangle must be less than half the area of the large square, and so less than the inner square in the first figure.

You might need to work on the two paragraphs of reasoning above with some pieces of paper, before proceeding to work with the software versions.

Task 1.4.1b Square and Rectangle in Dynamic Geometry Software

Use the dynamic geometry software file **'1c Square and Rectangle'** to explore both figures in Task 1.4.1 interactively.

Drag E between A and B and observe the relative sizes of the inner square and the inner rectangle.

Comment
Once you have selected a free point E on AB, points on other sides are constructed so as to complete the inner square and the inner rectangle.

In the previous task, you started with a square and investigated the behaviour of an interior square and rectangle drawn from a given point. The next task invites exploration of what happens if the initial square is replaced by a more general rectangle.

Task 1.4.2 Rectangles Within Rectangles

Given a rectangle and a point on one side, can you imagine drawing an interior rectangle with each vertex on an edge of the first rectangle? Is it possible to draw more than one interior rectangle from the point?

Comment
A sketch of the required figure is shown in Figure 1.4c.

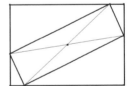

Figure 1.4c

It may seem intuitively obvious that the centre of the interior rectangle is also the centre of the original rectangle. Can you be sure? When trying to solve a problem in mathematics, it is standard practice to make use of a conjectured relationship without being absolutely sure of its validity to see where it takes you. Later you can go back and try to justify the conjecture. In this case, if you assume the centres are indeed the same, you could *start* with the centre in order to draw the interior rectangle.

The concept of 'centre' suggests a circle and this in turn suggests that you could construct the rectangle by constructing its diagonals. If you do this, you may find that you are surprised by the result. Try it and see. It is often the case in mathematics, that a solution gives more insight to a problem than you expected. In general it is possible to construct more than one interior rectangle from a given point. Use the interactive file **'1d Interior Rectangles1'** to investigate this further.

One of the things you may notice when dragging the point P in the interactive diagram is that *all* points on the shorter side of the rectangle can generate interior rectangles. Thus it is possible to draw an infinite number of interior rectangles. What happens to the vertices of the interior rectangles that lie on the longer side of the outside rectangle? It is not always possible to draw an interior rectangle if you start from any point on the longer side.

Since it is possible to construct an interior rectangle from *any* point on the shorter side of the original rectangle, this process can be repeated, as in the next task.

Task 1.4.3 Sequence of Rectangles

As before, imagine drawing a rectangle inside a given rectangle. Now imagine drawing another rectangle inside this one and continuing the process. What conclusion can you draw about this sequence of rectangles?

Open the interactive file '**1e Interior Rectangles2**' and drag the point A until your starting rectangle is a square as shown in Figure 1.4d. This is a very special case.

Now watch what happens when you drag the point A by a very small amount. The effect is quite dramatic and illustrates how the ratio of length to width varies for the interior rectangles.

Figure 1.4d

Comment

This task certainly lends itself to investigating with dynamic geometry software.

In the case of the square, the ratio of length to width of inner squares remains 1:1 (that is, is invariant). However, you will have found that the ratio is certainly not invariant when you start with a rectangle that is not a square.

When you consider all possible ratios of length to width for your starting rectangle, you are effectively looking at a range from 1:1 to n:1, where n can be as large as you like.

Reflection 1.4

What has been varied so as to provide you with experience of discerning details, recognising relationships, perceiving properties and reasoning with properties?

1.5 PEDAGOGIC PERSPECTIVES

This first chapter has introduced the mathematical themes that are taken up in succeeding chapters: invariance, language and points of view, reasoning, visualising and representing. Furthermore, it has introduced a number of pedagogic constructs and strategies that are of direct use in the classroom.

Pedagogic Constructs

The chapter began by suggesting that learners and teachers do not always stress or discern the same details. For example, Tom Bates (1979) reports that some primary school children will claim that triangle A in Figure 1.5 is not equilateral, but that triangle B is.

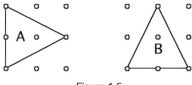

Figure 1.5

On rotating the board through 90° so that A has the same orientation that B had before, reactions have included: 'Oh, now it is!' What is the learner stressing, what is taken as invariant and what is seen as a permissible change? Simply subjecting learners to variation may not be sufficient for them to become aware of what the teacher sees as invariant, what they are permitted to change, and in what way.

In order to learn from experience, it is not usually sufficient just to have experiences. Some sort of reflection is at least beneficial, and usually necessary. In the reflection tasks in this chapter you were sometimes asked to look back and consider what role was played by 'invariance in the midst of change' – what sorts of things were permitted to vary, and in what ways, and what sorts of things remained invariant. Becoming sensitised to notice invariance in any mathematical task is good preparation for getting the most out of any task and the activity that arises from it.

Reference has been made to the fact that different people discern different details in geometrical figures (and in situations generally). Sometimes learners focus on recognising relationships (two angles equal, two sides equal, two shapes the same), sometimes on perceiving properties (isosceles means two angles equal and two corresponding sides equal) and sometimes on using properties to reason (since two sides are equal, two angles are equal, which means …). Noticing yourself focusing on different details at different times sensitises you to notice when learners are attending differently and so to adjust your teaching accordingly.

Pedagogic Strategies

The most effective pedagogic strategies for promoting learning are based around provoking learners into using their natural powers to make sense of the world, listening to what learners have to say as they do this and engaging in discussion with them. In short, effective teaching involves being mathematical with, and in front of, learners. A variety of strategies have been illustrated already in this chapter.

You may have noticed that instead of the usual 'can you find … ?' or 'find a … ', many of the tasks asked 'in how many ways can you … ?' Studies of classrooms around the world have revealed that where learners' attention is directed to multiple methods, they engage more fully, more creatively and more readily than when just asked to find an answer. In this text, the number of questions asked of you, the reader, has been restricted because it is not possible to modify the text on the basis of what you say and think. But the more you stop and think, question and challenge, the more you will get from doing the tasks presented in the text.

Many of the tasks in this chapter have been deliberately ambiguous. They have required you to recognise and make choices. This is especially true in counting tasks where what 'counts as the same' has to be decided by the person counting. Young learners may at first be frustrated by ambiguity, preferring to be told 'what is permitted', but once they become used to it, they tend to value it.

Asking learners 'what is the same and what different about' two or more objects or two or more figures is often fruitful, for it reveals what they are discerning and the relationships they are recognising. By hearing what others are saying they can see, learners have the opportunity to expand their own perceptions.

Dragging in Dynamic Geometry Software

Some of the interactive files you have opened have illustrated the power of the dragging tool in dynamic geometry software.

A construction can be explored by dragging relatively free elements in order to seek relationships in the figure that remain invariant. These can then be treated as properties and checked in other instances. To take a very simple example, learners might be asked to drag the vertices of a triangle, measure the interior angles and find the sum. Dragging the vertices of the triangle to different positions is likely to lead to a conjecture that the sum of the angles is invariant and is 180°. This is an empirical result, based on the large number of examples that one can quickly generate using dynamic geometry software. It may be intuitively convincing, but it cannot rule out the possibility of some extreme counter-example.

To be certain the conjecture is valid, you need to make use of underlying structural relationships by reasoning on the basis of agreed and accepted properties. Searching for structural relationships as the basis for reasons to explain why some relationship must always hold, and articulating the conditions under which the invariance is preserved, are fundamental to becoming a competent citizen as well as to appreciating mathematics (Bishop 1991).

Although dynamic geometry software is a powerful medium through which to encounter geometric relationships and properties, well-established principles recommend exposure of learners to physical objects. Handling physical objects (shapes, geoboards, geostrips, Polydron©, and so on) is an invaluable part of learning because it links physical movement and touch with geometric shapes rather than relying exclusively on sight and hand–eye co-ordination when using software.

Reflection 1.5

What struck you about the work in this chapter?

What do the following terms mean to you currently? Try telling someone else, or writing down something about them:

congruence, similarity, counter-example, construction, corresponding angles, properties.

What aspects of the tasks gave you some pleasure? What happened inside you? How might that happen for your learners?

2 Language and Points of View

This chapter shifts focus from invariance to the role of language for naming properties. This is a step on the way to developing reasoning on the basis of properties which is the subject of Chapter 3.

Throughout the previous chapter, the discussion involved the use of terms or descriptions that could be termed 'mathematical' in order to discern and identify different shapes, recognise relationships between the shapes and consider the behaviour of shapes under additional constraints. However, most of the terms were used with an implicit assumption that they are well understood and well defined.

2.1 DISCERNING AND STRESSING

There is a great deal of literature on the nature of mathematical discourse, and particularly on how children come to an understanding of mathematical language and reasoning, and the difficulties they experience. To take just one example of difficulties that arise, there are many words in mathematics that are also used in everyday language but with a different meaning; sometimes radically different, sometimes rather close. An amusing, but nevertheless illuminating, example is when a child describes the first angle shown in Figure 2.1a as a 'right angle' (having just been taught the term), then describes the second angle as a 'left angle' or the third angle as a 'wrong angle'.

Figure 2.1a

The child is discerning and stressing more than the adult. Another example is the word 'base', which has a number of different meanings within the world of mathematics itself (in area formulae for certain shapes, in vector geometry, in number representation) as well as its different everyday meanings. Moreover, the term 'base' when used in area formulae is a *relational* concept: its meaning depends on its relation to something else. There are many such words in mathematical vocabulary, perhaps not surprisingly, since mathematics is primarily concerned with relationships. For example, to describe a line as 'a perpendicular line' is not meaningful in itself. It must be perpendicular to something. To return to the term 'base', when using the formula 'Area of triangle $= \frac{1}{2}$ Base $\times$ Height' it is understood that the base and height referred to are a 'corresponding pair': for any side that is taken as base, there is a corresponding height. This can be a source of difficulty for learners for whom the concept of base is firmly rooted in its everyday connotation of being the 'bottom' of something. So the diagram on the left in Figure 2.1b apparently presents no problem, whereas the one on the right is not seen as including a base-height pair.

Figure 2.1b

Some of the important relational concepts encountered in Chapter 1 included congruent, similar, and parallel. Terms for shapes such as hexagon, quadrilateral and parallelogram involve relational concepts in their definitions. Even a seemingly obvious concept like 'equal' may need to be interpreted in terms of the context. For example, when developing the concept of fractions, the 'whole' is often referred to as being divided into 'equal parts'. This sounds innocuous enough, but consider the two diagrams in Figure 2.1c.

Figure 2.1c

The two shaded triangles in the rectangle on the left are both $\frac{1}{4}$ (of the area), but a young child is quite likely to argue that the two parts are not equal (identical in shape), so how can they both be $\frac{1}{4}$? They are *not* equal in the sense of being congruent. The child might eventually be convinced by the diagram on the right, which shows that each triangle is $\frac{2}{8}$, and all the small triangles *are* congruent. In what sense then, are the two triangles on the left equal? Clearly, there is an implicit assumption that the equality refers to equal areas, rather than congruent shapes. Meaning depends on context. When a learner says something that seems wrong to the adult, it may be that the learner is stressing something that the adult is ignoring, and vice versa.

This chapter looks more closely at some of these concepts. In particular, having investigated the behaviour of shapes under given constraints, you need to understand *why* they behave in that way. And this in turn means identifying the *properties* of the shapes. As before, many such explorations in geometry can be greatly facilitated by using dynamic geometry software.

Reflection 2.1

Think of some geometric words that have a different meaning in everyday life. What commonalities are there between the geometric meaning and the everyday meaning?

2.2 CONGRUENCE

You came across many examples of congruence in the previous chapter but what exactly is meant by 'congruence', and how, beyond the purely perceptual, can you convince yourself that two shapes are actually congruent? A simple physical interpretation would be to say that if you cut out two shapes and you can place one on top of the other so that they overlap exactly, then the two shapes are congruent. In practice, this is not always feasible as a form of reasoning.

Task 2.2.1 Corresponding

Try to be precise about the meaning of 'corresponding' in the statement that 'two polygons are congruent if corresponding angles and sides are equal'.

Be critical of your definition. Get someone else to see if they can make sense of it.

Comment
Did you think to specify that, for example, if two successive sides of one polygon correspond to two successive sides of the other, then the angles formed also have to correspond? Or perhaps it doesn't matter? This is the sort of question that is taken up below.

Do you need to check that every pair of edges, taken in the corresponding order, all need to be equal, in order to be sure of congruence? Is the same true for angles? And would checking edge lengths or angles alone be sufficient? In the case of triangles, some thought will show that if you know that two pairs of angles are equal there is no need to check the third pair; they must also be equal since the sum of the angles is 180° in every triangle. So the important question arises, 'what is a minimum set of conditions for determining congruence?' To simplify matters, it is helpful to concentrate on triangles to begin with.

Uniqueness

Congruence is a property involving relationships between two objects; uniqueness is a property of a collection of information about a shape. What information about a shape determines that shape uniquely?

Task 2.2.2 Reproducing a Triangle

Someone has drawn a triangle and measured all its sides and angles. What information do you need to get from them in order to reproduce an exact copy of the triangle? How many different types of information will suffice?

Comment

The focus of this task is on kinds of information about the triangle that are needed in order to draw a copy. Each time you ask for a measurement, or use a ruler, protractor or pair of compasses to copy a length or an angle, you are using information.

Just as with previous tasks in this book that have used the phrase 'how many different … ?', you need to ask what is meant by 'different'. Imagine the following incident:

> The teacher asks the group: 'What information do you require in order to draw exactly the same triangle as one that I have already drawn?' The triangle is not shown to the group but, in order to communicate efficiently, a 'generic' triangle labelled ABC (as shown on the left in Figure 2.2a) is drawn on the board. The teacher points out that his/her triangle may of course look like the one on the right.

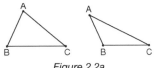

Figure 2.2a

> The teacher continues: 'You can now ask me for an item of data (for example, length of BC, angle B and so on) and after each item is given, we will check to see whether you have enough information to draw my triangle'.

By working in this way, checking at each stage, the idea of minimum conditions arises naturally. When a successful construction is completed, the process can be repeated for another triangle but with the added proviso that the same set of data cannot be asked for. At this stage, one point that will almost certainly arise and be discussed is whether, say, asking for AB, BC and ∠B is *effectively* different from AC, BC and ∠C.

Notice how attention shifts from specific information (particular sides and angle) to relationships (two sides and the included angle). This leads to the idea of *classifying* the condition, for example, in the form Side-Angle-Side or more succinctly as SAS. Some interesting aspects emerge from this activity. First, it is likely that the conditions SAS, ASA and SSS will be discovered by learners, often in that order. These three conditions can be nicely illustrated and reinforced with a visual aid such as 'geostrip'. Second, one of the 'standard' textbook conditions for congruence is referred to as RHS (Right-angle, Hypotenuse and Side) but this does *not* arise naturally from this activity. A third aspect is illustrated by the following task.

Task 2.2.3a An Ambiguous Condition

Referring to the generic diagram of a triangle ABC from the previous section, construct the two triangles with the information given below:

(i) $\angle B = 45°$, BC = 5 cm, AC = 6 cm.

(ii) $\angle B = 45°$, BC = 5 cm, AC = 4 cm.

(iii) $\angle B = 45°$, BC = 5 cm, AC = 3 cm.

What is the same and what is different about the three cases? Construct some other similar examples for yourself. What conclusions can you draw?

Comment
You probably found no problem in drawing a triangle for the first example, which might lead you to conjecture that the condition SSA is an 'apparently' legitimate condition. But the second example may give you cause for doubt. Did you notice that there are two triangles meeting the condition? The third example is impossible for any triangle.

The key to whether or not a triangle is defined depends on the relationship between AC, BC and the specified angle at B. The condition for uniqueness depends on the relationship between AC and BC.

Task 2.2.3b SSA, RHS and Dragging

Open the interactive file '**2a SSA Condition**' and investigate what happens when the lengths of the two sides vary. Then change angle B and vary the sides again. In particular, focus on what happens when $\angle B$ is a right-angle and when it is obtuse.

Comment
First, keep the angle constant and just vary the relative lengths of the sides. You will find that sometimes no triangle can be drawn. Determine the condition for this. Then try keeping the relative lengths of the sides constant and just vary the angle.

When trying to construct a triangle from *arbitrary* data, it is quite likely that you will choose some combinations from which it is impossible to construct a triangle. A simple example of this is the case of the condition SSS. Drawing a triangle with sides 10 cm, 4 cm and 3 cm is impossible, since $10 > 4 + 3$. This is an instance of the famous 'triangle inequality' that states a necessary condition for three lengths to be the sides of a triangle. It can be demonstrated very effectively with something like 'geostrip' or in dynamic geometry software. However, in the case of the group activity described above, this situation cannot arise because the data provided comes from a genuine triangle.

From your investigations, you will have seen that when AC < BC, two different triangles may be possible. This is why the condition SSA is described as ambiguous and why it is not traditionally accepted as a congruence condition. However, it is valid to ask: 'when *does* SSA give a unique triangle?' In general, this is when AC ≥ BC, because a circle centred at C with radius AC ≥ BC will necessarily cut the side BA just once on the correct side of B (so the angle B is correct).

This condition can be expressed even more generally by describing AC as the side opposite the given angle. If you can *guarantee* that the opposite side is longer than the adjacent side in SSA, then a unique triangle can be drawn. You may have noticed that the congruence condition RHS is a *special case* of SSA whereby you can guarantee that a unique triangle is defined because the hypotenuse is always the longest side. Now, just as a right-angle guarantees that AC > BC, similarly ∠B > 90° (that is, obtuse) also guarantees this fact. So it would be perfectly legitimate to classify SSA as a condition of congruence provided the equal angles are ≥ 90°. SSS, SAS and ASA determine unique triangles, while SSA determines a unique triangle if the side opposite the specified angle is the longer of the two sides.

Having investigated congruence conditions for triangles, it seems natural to extend this to quadrilaterals and other polygons. The uniqueness approach could be taken again, that is, trying to identify the minimum conditions necessary for determining a given quadrilateral.

Task 2.2.4 Congruence Conditions for Quadrilaterals

Open the file '**2b Quad Conditions**' and investigate the cases of four given sides and four sides plus one angle. What conclusions do you come to? How many pieces of information are required for a set of conditions?

Comment

In trying to find a set of congruence conditions, it is natural to try to define them in terms of the sides and angles of the quadrilateral, in the same way as for triangles. However, a quadrilateral can be divided into two triangles by a diagonal. Can this provide you with an alternative way of describing the conditions?

You will find that having four given sides does not provide enough information to construct a unique quadrilateral. Again, this can be illustrated physically using geostrip. (Heather McLeay, 2002, has written an interesting discussion comparing a geostrip approach with using dynamic geometry software.) With three strips hinged together the triangle formed is rigid, which corresponds to the SSS condition of uniqueness. This is also the reason why triangular structures are common in engineering projects such as bridges. Four strips hinged together form a quadrilateral that is floppy and sometimes can even be deformed from a convex shape to a concave one, as the interactive file demonstrates. It seems natural to suppose that if you also added an angle to the set of known data, this would make the quadrilateral rigid. You may have been surprised to find that this is not so. In an analogous way to the SSA condition for triangles, there is ambiguity because sometimes two different quadrilaterals can be drawn meeting the same conditions.

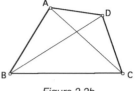

By considering the quadrilateral divided into two triangles, use the triangle conditions to help you. For example, consider the diagram in Figure 2.2b.

Figure 2.2b

If you focus on the triangles ABC and BDC, you know that each triangle will be uniquely determined by the SAS condition. That is, {AB, ∠B, BC} determine triangle ABC, and {BC, ∠C, CD} determine triangle BDC. Putting these together you would have a uniqueness (or congruence) condition for quadrilaterals that can be expressed as SASAS. By considering other triangle conditions for ABC and BDC, or indeed by considering other triangles, you could produce other sets of conditions for quadrilaterals. In fact, there is no reason why you should not introduce the diagonals themselves into your conditions.

Notice the use of labels to direct attention. It is conventional to use capitals for vertices and lower case for sides, and to label the vertices sequentially round the figure.

Focusing on triangles is useful in practical, real-life situations such as triangulation in surveying, though surveyors include extra information as a check on their measurements. The task of finding congruence conditions for quadrilaterals leads to appreciating the fundamental nature of triangles in relation to congruence and, indeed, to understand why it is not really necessary to memorise a list of specific conditions for other polygons.

Why, ultimately, do you need to know about conditions for congruence? An answer has already been suggested in the last section of Chapter 1: the need to explain, to justify, to convince and to deduce. These processes are fundamental to mathematical activity and, in the context of geometry, they often depend on being able to argue that certain given relationships imply that others will hold true. Congruence, and similarity, will often form the bedrock of such arguments.

Properties of Shapes

Shapes have features (sides, angles, and perhaps chords) but also perimeter and area, as well as colour, thickness and orientation. Some of these are mathematical, and some are not. One focused approach to investigating properties of shapes might be to start with a definition, construct the shape based on that definition and then explore other relationships within the shape. To take an example, a parallelogram is, by definition, a quadrilateral with its pairs of opposite sides parallel.

Task 2.2.5 Properties of a Parallelogram

Using the definition given above, sketch a parallelogram. Find as many properties of this shape as you can by measuring sides, angles, diagonals and so on. Check that the properties you find are invariant when you sketch a different parallelogram.

Comment
You can also explore parallelograms by opening the interactive file '**2c Polygon Properties**'. When you drag your parallelogram to different positions you may form some special parallelograms. What are they and what special properties do they have? Using the file you can also explore properties of other special quadrilaterals like trapeziums and kites. On each page of this file, the basic definition used to construct the shape is described.

Investigating parallelograms leads to a list of properties that probably include the following: opposite sides are equal, opposite angles are equal, the diagonals bisect each other. There are two interesting follow-ups to the process of compiling such a list. The first is: how sure can you be that these properties hold true for *all* parallelograms? The list is the result of a set of confirming instances and is therefore based on *inductive* or *empirical* reasoning. Even though dragging enables you to generate what seems like a very large number of such instances, and which may seem extremely convincing, it does not alter the fact that only a finite number of cases have been tested. The conclusions remain at the level of conjectures. However, using your knowledge of congruent triangles, you could reason *deductively* to justify conjectures, by making use of the definition of a parallelogram, and properties of parallel lines. For example, consider the diagram in Figure 2.2c.

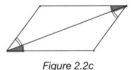

Figure 2.2c

Using properties of parallel lines and transversals, two pairs of equal angles can be identified in the two triangles. This shows that the triangles are similar. But the two triangles also share a common corresponding side. Therefore, the condition ASA can be used to show that the triangles are in fact congruent, or you can reason that one is a scaled copy of the other, and a pair of corresponding sides is equal, so the scale factor must be 1. From this it follows that corresponding pairs of opposite sides of the parallelogram must always be equal.

In the same way, for example, by introducing the other diagonal to the drawing and creating more triangles, more properties can be deduced. The same process can be applied to other quadrilaterals. Trying to justify the generality of properties you have found by using deductive reasoning involves taking stated properties as the starting point, and using known and agreed facts to produce a chain of reasoning that ends up with the conjecture. This is *reasoning on the basis of properties*. If, for example, you start with the property that the diagonals bisect each other, you cannot suddenly say that opposite sides are parallel (because you know or believe the figure is a parallelogram). You have to use only the properties you start with to deduce consequences.

Another follow-up to this activity is to take one of the properties you have found and to ask yourself a question like the following: 'If I draw (or make with geostrip) a quadrilateral with both pairs of opposite sides equal, will it be a parallelogram?' If you find that it is indeed a parallelogram, then this property gives you a different way to construct the figure. Starting from different properties and trying to deduce others is very much like adopting different points of view, and then trying to see if they lead to the same conclusions.

Paulus Gerdus (1988) describes how Mozambican farmers construct the rectangular base of their houses. Two ropes of equal length are knotted together at their mid-points. A bamboo stick whose length is the desired width of the house is laid down and two ends of the ropes are attached. The ropes are then pulled taut to determine the other two vertices of the rectangle. (See Figure 2.2d.) This method is also used by carpenters in Europe.

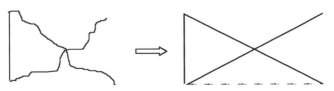

Figure 2.2d

Task 2.2.6 Constructing a Rectangle

Does the Mozambican construction always give a rectangle? Why? What properties of rectangles are being used?

Comment
This is a striking example of using the properties of a shape to devise a practical construction. Notice that, although most people would immediately think about right angles in relation to a rectangle, this method does not explicitly construct right angles.

Clearly, the property that is being used here is that the diagonals of a rectangle are equal and bisect each other. The question is, if a quadrilateral is constructed with this property, does it ensure that the quadrilateral will be a rectangle? Investigate this using dynamic geometry software and also try to argue deductively. The key feature that you must demonstrate is that the angles are right angles.

It is a valuable exercise to find alternative constructions for well-known special polygons such as a square, rhombus or kite. When you try to construct a particular shape, you find that the concept of minimum conditions appears quite naturally. For example, you might have a definition of a square in your mind as a quadrilateral with all four sides equal and all four angles right angles. It is undoubtedly true that a square possesses all these properties. But when you construct it, do you actually need to use all these facts? In practice, you will find that the square is completed before you finish using them all. For example, the sequence in Figure 2.2e illustrates one possible way of constructing a square.

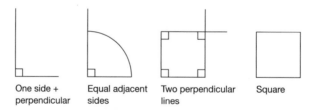

One side +   Equal adjacent   Two perpendicular   Square
perpendicular   sides   lines

Figure 2.2e

This may not be the 'best' way to construct a square (what could 'best' mean?) but it works. The point is that at each stage another known property of a square is being incorporated. Notice, too, that when the two perpendicular lines are drawn in the third stage, you could just as easily draw two lines that are parallel to the adjacent sides already drawn. When constructing using dynamic geometry software, you can check how successful your construction is by seeing if it is 'robust'. That is, when you drag the independent elements of your construction, the shape will vary in size and orientation but the intended shape remains.

The process of finding alternative minimum conditions is known as *characterising*, because the different sets of conditions each characterise the shape being studied, in the sense of providing properties from which all the other properties can be deduced.

Paper Folding

The art of origami has a very long tradition in Japan and is now a popular activity throughout the world. The word itself literally means 'folded paper' in Japanese. It is not the intention here to delve into the many complex models that can be created. However, if you open up even a simple model, you see a pattern of geometric shapes defined by creases. The following task focuses on creating shapes from creases.

Task 2.2.7 Folding Polygons

For each of the following polygons, take a sheet of paper and try to construct the shape by the creases produced through folding.

   isosceles triangle,   right angled triangle,   equilateral triangle

   rhombus,   rectangle,   square,   kite,   parallelogram

What makes one harder than another?

Comment

If you use a sheet of A4 paper, then you already have a rectangle that could be used to help fold some of the other figures. But it is more challenging to try folding without the help of already given sides and angles. On the other hand, most origami activities start with a square piece of paper and make use of the sides and angles.

When you are thinking about how to fold a shape, you almost certainly consider the properties of that shape. It is likely that you had to do quite a bit of experimenting because the required shape does not appear until you unfold the paper. You will probably have found that certain properties are more helpful than others in the context of folding. For example, right angles are easy to produce by folding and therefore, if the shape has any right-angled properties in it, this can be used to help you. Symmetry plays a big part. Another feature of paper folding is that when the paper is unfolded many repeats of a shape may appear; you may be surprised by just how complex a pattern emerges. Because of this, you almost certainly found that in solving one of the problems in this task you have in fact solved others too. A good example of this is the rhombus where the creases probably also gave you an isosceles triangle and a right-angled triangle.

In some cases, in order to make lengths equal, you may have produced a fold by placing one point on top of another. What is interesting here is that the fold, which will become an axis of symmetry, is a *result* of making two points coincide.

Task 2.2.8 Folding a Square in Half

Take a square sheet of paper and fold it so that the shape produced is half the area of the original square. How many different shapes can you make that have this property? Can one of your final shapes also be a square? Can it be a parallelogram?

Comment

You will quickly find one or two simple ways to do this. In order to find more possibilities you could try combining some of the 'basic' methods you have found. (A useful reference is Franco, 1999). Think about how you can be sure that your final shape is indeed half the area of the original.

The comment about comparing areas may have reminded you of Task 1.4.1 from Chapter 1. There, the focus was on whether the folded parts of the square overlapped or left some uncovered space. Clearly, if they fit together exactly (without overlap or space) then you can be sure that the final figure is precisely one-half. However, you may find that you have produced some overlap *and* some uncovered space. Then you have to convince yourself that the amount of overlap is *equal* to the amount of space. This may not always be easy. Another approach is to consider the area *outside* the new shape formed (that is, imagining the sides of the original square are still present).

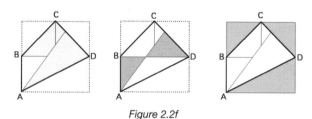

Figure 2.2f

For example, suppose the quadrilateral ABCD in Figure 2.2f is formed by folding as shown on the left. The two shaded triangles in the middle diagram illustrate the overlap and the uncovered areas. Take a moment to see why they must be equal in area. The third diagram illustrates the area outside ABCD. Take a moment to see why they, too, must account for half the area.

Reflection 2.2

What is the same and what is different about thinking and reasoning about the properties of shapes from a 'conditions for uniqueness' point of view, and a 'paper folding' point of view?

2.3 SIMILARITY AND RATIO

In many of the tasks in Chapter 1 you encountered similar shapes. In most cases the similarity was 'perceptually obvious' but, arising from Task 1.2.3, you were asked to determine whether two rectangles were similar in a situation where it was *not* obvious. What test did you apply? An intuitive description of similarity is that the *shapes* are identical (which implies corresponding angles are equal) but their *sizes* may be different. Another way to put this is to use a transformation term and say that one is an *enlargement* of the other. The implication of the first description is that the ratio of any pair of lengths *within* one shape will be the same as the corresponding pair in the other shape. The implication of the second description is that the ratio of a corresponding pair of lengths between the two shapes will be the same as any other pair of corresponding lengths *between* the shapes. Mathematically, these are in fact equivalent, but they represent two different ways of looking at similarity. For example, consider the two rectangles shown in Figure 2.3a.

Figure 2.3a

Looking at the ratios *within* the rectangles, the ratio of base: height for the smaller rectangle is 3:1 and for the larger one is 6:2 = 3:1. Looking at the ratios *between* corresponding sides of the rectangles, the heights are in the ratio 2:1 and the bases are in the ratio 6:3 = 2:1. If one of these lengths were unknown, but the rectangles were known to be similar, then either approach could be used to calculate the missing length.

Before considering ratios arising in triangles, here are some contexts that depend on ratios. Hans Freudenthal (1991) and his colleagues developed a very popular mini-project for mathematics classrooms in which the learners encounter a giant footprint 1.2 m long. They are then asked to find out all they can about the giant from this footprint.

This is a very open-ended task and in a classroom is likely to generate plenty of discussion about what kind of things can be determined and what kinds of assumptions need to be made. It provides access to a wide range of ideas and concepts, especially once the giant starts leaving messages for the children!

Clearly there are qualitative aspects of the giant about which nothing can be deduced, for example, colour of hair, whether the giant is fat or thin, and so on. To make quantitative estimates such as height, weight, hand-span or finger length, you have to make the assumption that the giant's body is similar in shape to an average human. This immediately raises questions about 'average' and, of course, a natural place for learners to start is with themselves. Here again, you can see the two different approaches concerning similarity that were identified earlier. You might measure your own footprint and compare it to your own height, then use this to estimate the giant's height. This uses internal ratios of similar figures. You could also find the ratio of the giant's footprint to your own and then multiply your own height by this value. This uses external ratios between the similar figures. Further mathematical issues develop when estimating the giant's weight, food intake, and so on. Another important concept about similar figures will emerge if you try to find the weight of the giant.

Consider now a real-life situation. When a film is shown on television a problem immediately arises because, although television screens and film screens are both rectangular, in fact they are not similar. The standard television screen has a ratio 4:3 (approximately 1.33:1), and a wide-screen television has a ratio of 16:9 (approximately 1.78:1). Film ratios vary considerably. Many films today have a 1.85:1 ratio but the Cinemascope ratio is about 2.35:1. The problem is, if the television and film ratios are different, how can a film be shown on the television screen? There are two 'solutions' to the problem. The first is to show the film in 'letterbox' format so that the film is seen in its original ratio but black strips appear at top and bottom of the television screen. So there is an unused part of the television screen. The second is to 'chop' the film so that it fits the television screen exactly. But this also results in unseen parts of the film (see Figure 2.3b).

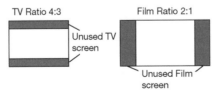

Figure 2.3b

Task 2.3.1 Films on Television

For convenience, assume you want to show a film whose ratio is 2:1 on a 4:3 television screen. Which method leaves unused or unseen the greater proportion of what is available?

Comment
It takes a while to find a point of view that makes sense of the situation. One approach is to choose the size of the television screen (using its ratio), then calculate the area of film seen (using the film ratio and the common width). Then find the unused area and hence the ratio of unused to whole. A similar calculation the other way round (changing your point of view) deals with the cropped film.
  The problem above gives particular values of the ratios. When you have found an answer, try the problem again with a different ratio for the film or the television, and then generalise.

Since the problem is concerned with the total unused or unseen in each case, where that portion is placed on the screen is irrelevant. It is easier, therefore, to do the calculations by shifting all the wastage to one part of the diagrams; that is, all the unused television screen at the top (or bottom) of the screen and all the unseen film at the right (or left) of the film. You may be surprised at the result. The result could just be a coincidence that arises from the particular ratios used. But mathematicians are very suspicious of coincidences! This is why it is worth investigating other cases and trying to come to a general conclusion. Most people do find the general result here quite surprising. And even if they can confirm it by say, using algebra (for example, taking the television ratio to be $a:b$ and the film ratio $c:d$) it is still difficult to see *why* it is true. This is the difference between a proof that logically confirms a hypothesis and one that provides *explanation*. As Michael De Villiers (1990) puts it, explanation aims to offer understanding and insight into why something is true, whereas a proof may be content with verifiable logical steps that give no insight.
  As with the television and film problem, you may be surprised by the answer to the next task.

Task 2.3.2 Circle and Squares

The diagram in Figure 2.3c shows a circle with one square inside and one square outside. Find the ratio of the areas of the two squares.

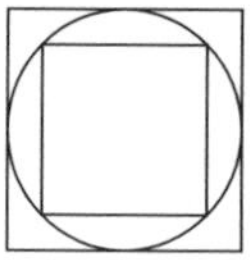

Comment
As an extension to this, consider a second circle drawn inside the inner square. What is the ratio of the areas of the two circles? Do you need to do any calculation to find the answer?

Figure 2.3c

You may have had the feeling that the relationship should be obvious just by looking at the diagram. However, if you solved the problem by drawing in some radii and then perhaps using Pythagoras' theorem, you may be convinced by the logic of the argument without appreciating geometrically why the relationship holds. An insight appears when you change your viewpoint from seeing the inner square as initially displayed, to seeing it as somewhere within the circle more convenient. Look at the interactive file '**2d Circle and Two Squares**'. No formal proof is given in this file but it does give real insight into the relationship. Writing out what you notice provides a stepping stone to geometrical reasoning.

If a second circle is added, as suggested in the comment above, then you could see the diagram as consisting of two similar figures. Each figure is a square with a circle inside it. Since the ratio of the areas for the two squares has already been established, it follows that the ratio for the two circles will be the same.

In his book *How to Solve It*, George Polya (1957) poses the problem illustrated in the next task.

Task 2.3.3 Polya's Square–Triangle Problem

Given a triangle, draw a square inside the triangle such that two of its vertices are on the base of the triangle and the other two vertices are on the other two sides of the triangle.

Comment

A possible solution to this problem uses the concept of similarity and arises naturally from a useful problem-solving strategy suggested by Polya; that is, to first simplify a problem by 'loosening' some of its constraints. The constraints of this problem are the requirements on the positions of the vertices of the square. Suppose you just want to draw a square with two vertices on the base of the triangle. Now add in one more constraint. Suppose you want to draw a square with two vertices on the base and one of its other vertices on one side of the triangle. You could draw a figure like the one on the left in Figure 2.3d by first drawing a vertical line AB, then constructing AD to be the same length as AB using a pair of compasses.

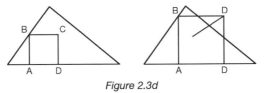

Figure 2.3d

Polya then considers drawing many such squares and observes how the point C varies. Polya first published his book in 1945 (although there have been many subsequent editions) long before all the advances of modern technology. But it is very striking here how his suggested strategy lends itself perfectly to exploration with dynamic geometry software. Imagine dragging point A (or whichever is the independent point that you started your construction from) and observe the path traced out by C. In fact, in most dynamic geometry software there is a named 'trace' function that enables you to see the path of a point. The diagram on the right in Figure 2.3d shows part of the trace of the point C from such software. Once you know the path, the task becomes one of constructing that path from given information (in this case, two positions of the square can be used to draw the line which intersects the side of the triangle). Then the vertex of the required square can be constructed as an intersection. The trace or path of a point is known as the 'locus' of the point. You will think about such ideas in more detail in Chapter 4.

Returning to the television and film problem, is there some way of demonstrating the relationship that will give some insight into why it is so? It seems that somehow it might help to combine the two diagrams in Figure 2.3b into a single diagram. This will be easier if the wasted areas are shifted, as suggested earlier. Look at the interactive file '2e Film and TV'. By dragging as indicated, you may be able to 'see' why the relationship holds. There are some similarities with Polya's square–triangle problem, in terms of points of view as ways of thinking.

Seeing connections between different problems in mathematics, and connections from one area of mathematics to another, is a fundamental aspect of thinking mathematically. It is also one of the reasons why mathematics is often exciting, challenging and even beautiful. This is why, throughout this book, you are encouraged to see these links and to make them for yourself. Properties of circles will be investigated in more depth in Chapter 3.

Reflection 2.3

Describe to yourself as many connections between similarity of shapes and ratios as you can.

2.4 MID-POINTS, BISECTORS, PARALLELS AND PERPENDICULARS

The focus so far has been mainly on investigating shapes and their properties, together with relationships between shapes like similarity and congruence. Having seen that concepts such as parallel lines and perpendicular seem rather significant, you could focus your attention on these concepts themselves and *introduce* them into shapes to see what happens. First, consider a line parallel to a side of a triangle.

Task 2.4.1 Parallel in a Triangle

Draw any triangle ABC. Mark a point D on AB and draw a line DE parallel to BC. (See Figure 2.4a.)

By imagining, or by actually dragging the point D in dynamic geometry software, and by dragging the vertices of the triangle ABC, investigate the ratios DE:BC, AD:AB, AE:AC. Explain the invariances you detect. Also consider the ratios AD:DB and AE:EC.

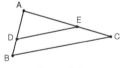

Figure 2.4a

Comment

There is an interactive file '2f Parallel' associated with this task. The first part of this task involves discerning similar embedded triangles (ADE and ABC) that should enable you to explain the results. However, for ratios such as AD:DB, it is not obvious what similar shapes AD and DB belong to. Explaining that result is therefore more challenging. If you are not sure about it, look at the interactive file '2g Triangle Intercepts'. The relationships are known as Thales' theorem, after a mathematician who lived in Greek Ionia around 600 BCE. His theorem shows that if DE is parallel to BC then AD:DB = AE:EC = DE:BC (and hence certain other ratios are also equal), and vice versa, that if AD:DB = AE:EC are equal (for example), then DE is parallel to BC.

A special case of Thales' theorem occurs when D and E are taken as the mid-points of AB and AC respectively, for then DE:BC = AD:AB = 1:2. This result is known as the Mid-point theorem. Chapter 3 will show how Thales' reasoning can be extended to develop further properties of triangles.

Many mathematical investigations arise from questions such as 'What happens if … ?'. Learners of all ages benefit from being encouraged to form the habit of asking such questions. For example:

What happens if you bisect the angles of a triangle?
What happens if you construct the three heights of a triangle?
Does the same thing happen with quadrilaterals?
The median of a triangle is a line that joins a vertex to the middle of the opposite side. What about medians for quadrilaterals? How are they defined?
What about special triangles and quadrilaterals?
What about other polygons?

Sometimes questions and ideas come to you arising out of an investigation or problem. For example, if you have tried the list of questions above, you may have been struck by how often the lines you drew in the triangle (for example, bisectors, perpendiculars) were concurrent. This might stimulate you to think about problems *starting* with an arbitrary point inside a triangle. What could you do with this point? Join it to the vertices perhaps?

Task 2.4.2 Point Inside a Triangle

Draw a triangle ABC and mark any point P inside the triangle. Join P to A, B and C and mark the mid-points of each segment. Join these mid-points to form a new triangle. Now imagine dragging point P around inside ABC. What happens?

Comment
After you have tried imagining, look at the interactive file '**2h Point in Triangle**'. Certain features may be apparent very quickly. Others may not be so obvious. Notice the area of the inside triangle. What do you find? Is it possible to explain everything that happens?

Sometimes a situation makes more sense when it is generalised. For example, you might ask yourself whether the 'same things happen' for quadrilaterals and other polygons? You started by marking a point *inside* the triangle, but when dragging in dynamic geometry software there is nothing to stop you moving the point outside to see what happens. By joining the mid-points, you have created parallel lines. Why not take *any* point on one of the segments and produce an inner triangle by drawing parallel lines. What happens then? Is it possible to find a position for P such that the trapeziums outside the inner triangle are all equal in area?

Although there is a great deal more to find out about triangles, quadrilaterals are also interesting!

Task 2.4.3 Mid-points of Quadrilaterals

Using dynamic geometry software, draw any quadrilateral and then join the mid-points of the sides. This gives a new quadrilateral inside the original one. What can you say about this interior quadrilateral?

Comment
When dealing with any general proposition concerning quadrilaterals it is always of interest to consider special cases too (after all, there are so many special quadrilaterals!). Think about a square, rectangle, parallelogram, kite and so on as your starting quadrilateral. What happens to the interior quadrilateral?

It probably did not take you long to recognise that the interior quadrilateral always appears to be a parallelogram. Of course, the special cases produce parallelograms too but, not surprisingly, most of these are also special parallelograms. Some of the examples you may have considered are shown in Figure 2.4b.

Figure 2.4b

There is an interesting 'duality' between the rectangle and rhombus: each seems to produce the other leading to the question 'Can a quadrilateral, other than a rhombus, produce an interior rectangle?' and vice versa. If you looked at the case of a kite or a symmetric trapezium you will already have answered this question.

This raises the question of what *determines* the type of parallelogram that is produced. And to answer this, you need to go back to the original claim that the interior quadrilateral is always a parallelogram. This remains as a conjecture unless you can *explain why* it is a parallelogram. Once you have understood this, you will probably be in a better position to explain the special cases as well. On the other hand, observation of the special cases might give you a clue to a general explanation. Both of these scenarios occur quite often in mathematics. That is, the particular may lead to a solution of the general, or a solution of the general may help in understanding the particular.

The mention of mid-points may remind you of mid-points in a triangle discussed earlier. The problem is that there are no triangles to be discerned. But alerted to the possibility of triangles, perhaps some can be manufactured by inserting some a segment or two as shown in Figure 2.4c.

By the Mid-point theorem, both AB and DC are parallel to and one half of the diagonal. Therefore they are parallel and equal to each other, and this in turn guarantees that ABCD is a parallelogram. To see that the other two sides of the parallelogram are equal and parallel you could use similar reasoning by inserting the other diagonal.

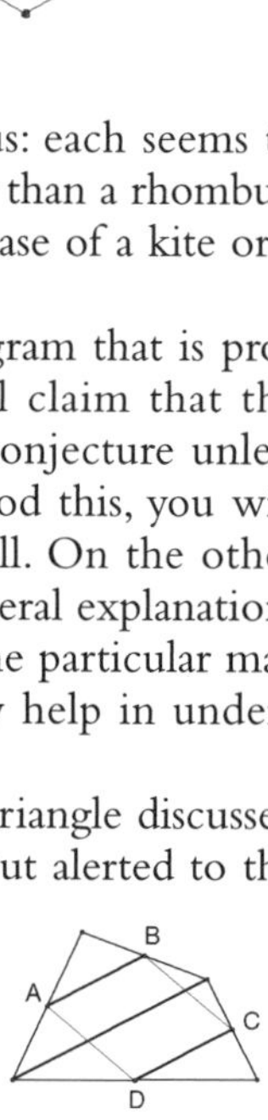

Figure 2.4c

Seeing the relationship between the sides of the parallelogram and the diagonals of the quadrilateral helps in understanding the results of the special quadrilaterals, but it also allows you to go further. You can use it to construct quadrilaterals with specified special mid-point parallelograms, and even to characterise those quadrilaterals that give rise to particular mid-point parallelograms. Michael De Villiers (1990) discusses the example of the interior rectangle in his interesting article about the different functions of proof.

While investigating Task 2.4.3, you may have dragged the vertices of the quadrilateral into a concave position and even into a 'cross-over' position as shown in Figure 2.4d. You may have been surprised to see that such dragging did not affect a parallelogram being formed but now you can understand why.

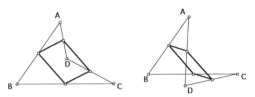

Figure 2.4d

In order to reason on the basis of properties, it is necessary to use mathematical language precisely, and to follow consequences. Sometimes it is necessary to insert or construct extra elements that enable you to discern sub-figures whose known relationships (for example, similar triangles and ratios and parallelism) you can use to make deductions about other relationships. Notice also that it is important at the end of a chain of reasoning to look back at what was invariant, and to clarify what it is that is allowed to change and in what ways.

> ### Reflection 2.4
>
> Look back over the reasoning given in this section and consider what particular features of language are involved in reasoning on the basis of properties. How has changing your point of view supported you in seeing why some relationship must always hold?

2.5 PEDAGOGIC PERSPECTIVES

The introduction to this chapter discussed some aspects of language in relation to mathematical concepts and relationships. Many of the tasks have encouraged you to think about these concepts and relationships in different ways and to try to come to an understanding of their properties. Such understanding is intimately connected with the language that is used to describe the ideas. This is not only a question of formal definitions. Indeed, in moving along the path towards understanding, formal definitions may be the least useful vehicle to use. As David Ausubel (1963) explains, one way to measure understanding is to see if the learner can express an idea in their own words rather than reciting a definition verbatim. There are two aspects in particular that have pedagogic implications and are therefore useful to focus on. One is the idea of 'negotiating meaning' and the other concerns what teachers can learn about children's understanding by *listening* to them talk about mathematical ideas. Both of these aspects once again illustrate the importance of discussion in the learning of mathematics.

Children's Descriptions

Getting children to talk about mathematical ideas and concepts usually requires some form of stimulus from the teacher, although sometimes it may be spontaneously initiated by the children themselves asking a question. The following example illustrates how the stimulus of asking for descriptions of geometric figures can give rise to important insights about children's understanding. This in turn could lead to valuable discussion about the concepts involved.

Some groups of children were asked to write a description of a number of geometric figures. They were then asked to write an alternative description for each figure. This was to emphasise that there is no single answer to the question; it is a matter of how *you* see it. Before reading further, write your own descriptions for the diagrams illustrated in Figure 2.5a, and get some children to write, or tell you, their descriptions.

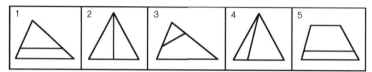

Figure 2.5a

Although there are no 'right' answers for this task, sometimes descriptions demonstrate some misunderstanding, or incomplete understanding, about certain concepts. To this extent, such a task can act as a diagnostic tool, in that it may uncover such problems. On the other hand, a perfectly good description does not necessarily imply complete understanding. Consider these two descriptions given by the same student:

This shows a triangle with a perpendicular to $\frac{1}{2}$ of the base (diagram 2).

This shows a triangle with a perpendicular to $\frac{1}{4}$ of the base (diagram 4).

What does this tell you about the student's concept of perpendicular? Another student gave the following identical description for diagrams 1 and 3.

One triangle is enlarged to another by a scale factor from the vertex.

The important thing to consider here is not just the responses and whether they seem interesting or not, but on how you would follow up on and discuss the responses. For example, how would you react to the following descriptions?

This shows two similar triangles, one enclosed inside the other (diagram 3).

Here are two similar trapeziums (diagram 5).

Ask yourself whether the first of these could be true. What would it require? If the second statement was supported by the argument 'It's just like diagram 1' how would you respond?

To give a flavour of the range of different responses produced by this task, here are some descriptions for diagram 1. In each case, consider how you would try to develop the subsequent discussion.

It is a triangle and a parallelogram.

It is a triangle with 3 sharp edges and 2 parallel lines.

Two triangles with same base and height of different measurement.

Two triangles with one same angle.

It has a double adjacent.

From a 3-d view, a triangle-based pyramid with the top part horizontally cut off.

In many of these cases, your first task would be to try to get the learner to explain his or her description a little more. Indeed, you may not be at all sure for some of the descriptions, just what is meant. However, as long as you do not try to *impose* your view, but rather take the student's description as a starting point for further thinking, the outcome is likely to be fruitful. In fact, this now leads to the first aspect mentioned earlier, namely, negotiating meaning.

Negotiating Meaning

Any form of genuine discussion involves the negotiation of meaning between participants. Unfortunately, many of the interactions that take place between teacher and students are not at all like this but, rather, take the form of 'transmission of knowledge' from one to the other or of 'guessing what is in the teacher's mind'. However, in using the idea of negotiation, it is not the case that everything is 'up for grabs' and that mathematical concepts can be interpreted in any way one wishes. The anarchy of Humpty Dumpty's 'Looking Glass' logic would not be very helpful!

> 'When *I* use a word,' Humpty Dumpty said, in rather a scornful tone, 'it means just what I choose it to mean – neither more nor less.'
>
> 'The question is,' said Alice, 'whether you *can* make words mean so many different things.'
>
> 'The question is,' said Humpty Dumpty, 'which is to be master – that's all.'
>
> (Lewis Carroll, *Through the Looking-Glass*)

The point here is that the very act of discussing, involving the sometimes tentative use of new words, is part of the process of arriving at meaning and understanding. And it is the teacher's duty to help children achieve this understanding for themselves, by starting from their initial position or by challenging them to think about a mathematical idea from a different viewpoint. You saw something of this in Chapter 1, with the discussion of exterior angles for concave polygons. Consider the following example, arising from a geometric problem in the classroom. The students had been asked to find the length of arc PAQ in terms of x, the circumference of the circle (Figure 2.5b).

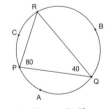

Figure 2.5b

Prior to this problem, the students had discovered that arc lengths are proportional to the angles at the centre, and to the angles at the circumference, subtended by the arc. They had also discovered the important relationship that the angle at the centre is twice the angle at the circumference. This is what the teacher was expecting them to use. However, one student produced the following solution, which puzzled the teacher.

$$\frac{\text{arc PAQ}}{x} = \frac{60°}{180°} = \frac{1}{3} \quad \text{Hence arc PAQ} = x/3$$

When discussing this solution, the student explained that, having found the angle subtended at the circumference by arc PAQ was 60°, he wanted to compare this to *the angle subtended at the circumference by the circumference*. He then argued that the circumference was composed of arcs PAQ, QBR and RCP and they subtended angles 60°,

80° and 40° respectively. The sum of these angles would be the required angle at the circumference subtended by the circumference.

The italicised phrase used by the student is probably one that a teacher would not use. What meaning can be attached to such a phrase? The student endowed it with his own meaning. By going with the learner's idea, a fruitful discussion, even a debate, could arise, for example: by considering other partitions of the circumference into arcs; by investigating the idea of adding the angles at the circumference even when they are at different points on the circumference; by comparing this situation to the angle subtended at the centre by the circumference; by taking a 'limit' approach. The interactive file '**2i Circumference**' takes this latter approach.

This chapter concludes with one more example that illustrates the process of negotiating meaning. It is not discussed in detail but is more of a stimulus for thought. Consider the concept of parallel lines. Railway lines are often given as a real-life example. Children may make the observation from this that railway lines sometimes curve. So can you have parallel curves? In particular, can you have parallel circles? If you go to a dictionary (for example, the *Oxford English Dictionary*) to resolve this question you would find the phrase 'continuously equidistant' used to define parallelism of lines, and the definition of line includes the phrase 'straight or curved'. Well, there would certainly be trouble if railway lines were *not* always the same distance apart! Does this mean that two concentric circles can be described as 'parallel' circles? Imagine the inner circle getting smaller and smaller until it becomes a point. Does this mean a circle can be described as a line parallel to a point? Perhaps all this sounds a little bizarre, but whenever definitions are introduced these are the kind of problems that must be faced. For example, what exactly is meant by *equidistant* in this context? How do you measure the distance between two straight lines or between two curves? You could follow up on the idea of concentric circles as parallel by considering a transversal and see if the same properties hold for this as for parallel straight lines. In fact, the concept of parallelism holds a very significant place in the history of mathematics. If you are interested to know more about this, type in the phrase 'Euclid's Parallel Postulate' into a web search engine.

Reflection 2.5

What struck you about the work in this chapter?

What do the following terms mean to you after working on this chapter? How has your understanding changed:

congruence, similarity, counter-example, construction, corresponding angles, properties?

3 Reasoning Based on Invariant Properties

This chapter is concerned with reasoning on the basis of properties, with a particular focus on the role of invariance. One of the activities that Alan Bishop (1991) identifies as playing a central role in enculturation of mathematical thinking is *design*, in the sense of acting on and manipulating the environment using mathematical knowledge. In geometric terms, design is essentially concerned with geometric constructions, and reasoning is central in validating geometrical construction.

3.1 GEOMETRIC REASONING

In terms of its geometrical content, this chapter will focus mainly on circles. Of course, circles have already featured a number of times in some of the tasks in previous chapters, but until now the focus has not been specifically on their fundamental properties and associated invariants. Although the design of buildings in the environment tends to be 'rectangular' in the developed world, in many cultures of the Third World the circle has been a more prominent design feature of dwellings.

However, in terms of the artefacts and machines people use daily, the circle dominates. Where would people be without wheels and gears? Not only is the circle of the greatest importance in the design of machines, it is also one of the few Euclidean shapes that can be observed easily in nature (that is, without the help of instruments such as microscopes). The sun and moon, the centres of flowers, cross-sections of tree trunks, and the iris of eyes are all apparently circular in shape. Circles have an aesthetically pleasing 'perfection' and many beautiful designs are based on circles and their intersections. It is also the most 'compact' of shapes in the mathematical sense that for a given perimeter it encloses the greatest area. Or, to put it another way, for a given area it is the shape with the least perimeter. Similar concepts hold true in three-dimensions, where a sphere is the most compact solid.

In order to illustrate the process of geometric reasoning in a quite difficult problem, consider the following result.

Given any triangle ABC, mark any point D on AB and draw a line DE parallel to BC as shown in Figure 3.1a. Join BE and CD and mark the intersection F. Then draw the line AF and extend it to intersect BC and label the point of intersection G. Then the line AFG is always a median of the triangle.

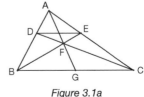

Figure 3.1a

How can it be proved that AG is always a median? That is, how can it be shown that no matter how the triangle changes its shape and no matter how D is positioned on AB, the construction described leads to G being the mid-point of BC?

First, consider the property of G being sought. For G to be the mid-point of BC, BG = GC. Notice how the property of AG being a median is transformed into the property of G being the mid-point, which is transformed into an equality BG = GC. This sequence of transformations is entirely based on definitions. Now there is something concrete to prove: an equality, which releases facts about congruence and similarity to be used.

The proof must use only properties given by the situation (DE being parallel to BC) and known facts about triangles such as congruence and similarity conditions. Now it is appropriate to examine the figure. Bear in mind that the 'figure' is not a single figure but *all* possible figures formed by changing the shape, size and position of the triangle ABC. Whatever properties this figure possesses must arise from the line DE being parallel to BC. Attention focuses on DE, and then looks for familiar figures incorporating DE, such as triangles. Then you can look for invariances associated with the fact that DE is parallel to BC. Wherever D is positioned, and DE drawn parallel to BC (and then BE and CD drawn), triangles are formed, which are similar, as shown in Figure 3.1b. Focus your attention on triangles ADE and ABC, ignoring other lines and points. Why must triangles ADE and ABC be similar?

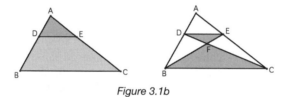

Figure 3.1b

Now shift your attention to triangles DFE and BFC, to see why they too must be similar. Thinking of the similarity between a pair of triangles in terms of a scale factor, you can write the scale factor between the first pair as BC/DE = k say. That is, BC = k.DE (and of course, AB = k.AD and AC = k.AE). But the scale factor for the second pair must also be k since DE and BC are corresponding lines for that pair of triangles (and here you also have CF = k.DF and BF = k.EF). The fact that both pairs of similar triangles have the same scale factor is a crucial feature of this situation.

What happens when the line AFG is drawn? As you can see in Figure 3.1c, the parallelism of DE and BC once again ensures that similar triangles are produced.

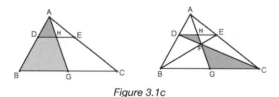

Figure 3.1c

The scale factors between triangles ADH and ABG and between DFH and CFG are each again k (provide your own reason!). This means that BG = k.DH and also that GC = k.DH. Therefore, you now have the desired conclusion that BG = GC.

Whenever you are trying to explain or provide reasons for something in geometry you need to focus on the (invariant) properties that are a result of the conditions of the problem or that arise from any constructions you introduce into the problem.

Reflection 3.1

How important for you is believing the result before you set out to prove it? Do you think it would have helped you to have experimented with an interactive file at some stage while you were working on the reasoning?

3.2 CIRCLE PROPERTIES

What is a circle? From a 'static' perspective, you can see a circle as a plane shape that is perfectly symmetrical about any line through its centre, or whose perimeter (circumference) is always equidistant from a fixed point (centre). Alternatively, you can see a circle from a 'dynamic' perspective and describe it as the path traced out by a moving point that is always a fixed distance from a fixed point. You can also think in terms of angles subtended by any diameter as being a right angle, or even angles subtended by any specific chord being the same. The list goes on. Circles have many different properties, but one must be chosen as the defining property, and then all the others need to be deduced from this.

This chapter investigates a few of the relationships and properties that emerge when you consider circles and lines intersecting each other in the form of chords and tangents, and circles intersecting with other circles. For most of the tasks in this chapter it is important to make use of dynamic geometry software.

Task 3.2.1 A Circle and a Line

Draw or imagine a circle and draw or imagine any line outside the circle. Translate the line (drag it in parallel fashion) across the circle. When it crosses the circle, construct the segment (that is, chord) AB as shown in Figure 3.2a. Construct or imagine the mid-point of AB and trace its position as the line moves right across the circle. Based on relationships you recognise, what properties of a circle can you conjecture?

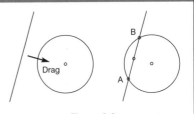

Figure 3.2a

Comment

This activity is based on a suggestion by James King (1996) in a useful book on learning geometry using dynamic geometry software. You will probably be able to make more than one conjecture. Notice that even though you have imagined, drawn or constructed in dynamic geometry software only one circle and one line, the situation is perfectly general and, so, relationships are going to be general properties, not specific to your circle and your line. So what is free to change is the circle and the slope or orientation of the line.

Having made some conjectures, the task is to explain or justify them. In fact, this section of the chapter is more concerned with discovering properties and relationships rather than rigorously explaining them. Dynamic geometry software can often give you a better 'feel' for properties than insight into ways of justifying them through geometrical reasoning. Nevertheless, most of the properties of circles can be justified by returning to the defining property of a circle, inserting radii, and observing that isosceles triangles are bound to occur due to the radius property.

The basic properties of chords and tangents are no doubt familiar to you, but by using the dragging facility in dynamic geometry software they become more accessible to learners' imagination. The trace of the mid-point in this task not only suggests something about the behaviour of chords, but also illustrates a fundamental property of tangents: the trace is a line perpendicular to the chord through the centre of the circle, and this remains true when A and B coincide to give a tangent instead of a chord. Notice that the definition of a tangent is 'a line touching the circle in one point' (or, if you like, two coincident points). Thus, the fact that it is perpendicular to the radius at its point of contact is a *property* of tangents. It is often mistakenly assumed that this is part of the definition. Here is another way to investigate the tangent property.

A secant is the technical name for a line *cutting* a curve in one or more points. Clearly, for a circle, a secant cuts it in two points.

Task 3.2.2 A Secant Becoming a Tangent

Construct or imagine a secant that crosses the circumference of a circle at points A and B. Join A and B to the centre O of the circle. Focus your attention on the angles AOB, OAB and OBA, and then drag point B along the circumference until it coincides with A. What are your conjectures?

Comment
It is important in constructing the secant that the two points defining it are on the circumference of the circle. This is quite different to the previous task where the initial points defining the line were independent of the circle.

This task also leads to the perpendicular property of tangents. However, precisely because it *is* a property, you should also be able to explain why it always holds, why it is an invariant.

Task 3.2.3 Intersecting and Touching Circles

Open the interactive file '**3a Touching Circles**' (see Figure 3.2b). Drag the centre of either circle so as to change its radius until A and B coincide. What appears to be the case about the line through A and B at this point? Describe how the point when A and B are coincident is related to the centres of the two circles.

Comment   *Figure 3.2b*
Notice what can change and still preserve the invariant relationship you found.

Task 3.2.1 involved a single line producing parallel chords. Imagine isolating two examples of chords as the line is dragged in parallel fashion across the circle. Joining the end points of two such chords produces a trapezium inside the circle, as shown in Figure 3.2c.

Figure 3.2c

Task 3.2.4 Trapeziums in a Circle

Draw a pair of parallel chords in a circle, and complete to form a trapezium and imagine dragging the parallel chords to different positions. Focus on the angles and the non-parallel sides. What conjectures do you come to? Justify them.

Comment
You may feel that there is hardly any need to justify your conclusions because they seem intuitively obvious. Certainly intuition is a powerful and valuable element in the process of doing mathematics but, unfortunately, it cannot always be relied on. Remember that one of the functions of proof is to try to explain why something is so, and this chapter is concerned with the reasoning process leading to such explanation.

You may have based your justification on a 'symmetry argument'. That is, you may claim that the diagram is obviously symmetric and this is why there are two pairs of equal angles in the trapezium and why the non-parallel sides are equal. Such a trapezium is called an *isosceles* trapezium but is also referred to as a *symmetric* trapezium. You are therefore in danger of circular reasoning here (forgive the pun!). That is, you may be saying that the sides and angles are equal *because* the trapezium is symmetric, but the assertion that the trapezium is symmetric may be based on the *assumption* that the angles and sides are equal.

Figure 3.2d

There are a number of ways that you might explain the conclusion deductively. One approach is to go back to a fundamental property of chords that you discovered in Task 3.2.1 (that is, a chord is perpendicular to the radius through its mid-point). Now consider the diagram in Figure 3.2d.

In Chapter 2 the idea of congruence conditions for quadrilaterals was discussed and the point was made that, in general, it is not necessary to memorise such conditions because they are built up from congruent triangles and, so, these can always be used instead. However, this seems a perfect opportunity to use a condition that was established in Task 2.2.4, that is, SASAS. If you consider the quadrilaterals AEFD and BEFC, you can immediately establish their congruence using this condition.

Having found that two parallel chords give rise to an isosceles trapezium, an interesting consequence emerges. Consider the isosceles trapezium in Figure 3.2e.

Figure 3.2e

Triangles ADC and BDC are congruent and, therefore, ∠ACD = ∠BDC. However, because AB and DC are parallel, it is also true that ∠ABD = ∠BDC. Therefore one can conclude that ∠ACD = ∠ABD. This is interesting because both angles are subtended by the same chord AD, but triangles ABD and ACD are neither congruent nor similar. Is this something that is specific to trapeziums inside a circle, or is some more general invariance at work?

Task 3.2.5 Angles Subtended by Chords

Open the interactive file '**3b Angle Chord**'. Drag point C around the whole circumference. What do you observe? Try different starting positions for the chord AB. What happens if AB is a diameter?

Comment
You will find that abrupt changes occur to ∠ACB as you drag C past A and B. You might then find it useful to draw two separate parts to your diagram and investigate the relationship between the two parts. For example, re-label the point at the circumference as C' when it is on the other side of AB. What is the relationship between ∠ACB and ∠AC'B? What can you therefore say about the internal angles of any quadrilateral drawn in any circle?

Some very significant relationships concerning angles subtended by chords can be discovered in this task. Apart from what you have already observed, join A and B to the centre O of the circle and measure ∠AOB. The fact that it is twice ∠ACB is the most significant relationship of this situation. You may remember that this was already mentioned in the discussion at the end of Chapter 2. The other relationships that you will have found can all be derived from this property. Try to explain them all by using this property. For example, when AB was a diameter, what was the angle subtended at the centre? So what can be said about the angle subtended at the circumference by a diameter?

Task 3.2.6 Constructing Tangents

You have already considered constructing a tangent to a circle at a given point on its circumference. Using what you have learned from the previous task, devise a method for constructing a tangent to a circle from a point outside the circle.

Comment
A useful strategy for solving construction problems, which has been mentioned before, is to draw a sketch of the required figure and think about the properties it must have. (See Figure 3.2f.) Introduce elements that contribute to displaying useful sub-figures such as parallel lines and isosceles triangles. Then you can focus on how to introduce those properties into the given situation.

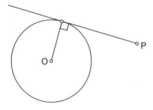

Figure 3.2f

All you have to start with is a circle and a point outside it. When the tangent is constructed, you know that it must be perpendicular to the radius at the point of contact. How can you construct a right angle standing on the points O and P? (Think back to the diameter situation in the previous task). One of the things that you will have immediately realised when you did this construction is that there are two possible solutions. Notice that the lengths of the two tangents (from P to the points of contact) are equal. This, too, can be proved by reasoning from the property that tangents are perpendicular to the radius at their point of tangency.

> **Reflection 3.2**
>
> Make a list of properties of circles. See if you can prove that these properties characterise circles: that is, starting from any one of them, all the others can be deduced.
>
> With each property, articulate to yourself what can change and still that property holds.

3.3 CIRCLES, TRIANGLES AND QUADRILATERALS

While investigating the properties of circles in the previous tasks, you have also been dealing with the relationships between circles, triangles and quadrilaterals. This section focuses on these relationships and looks at some of the implications and applications.

> **Task 3.3.1 An Archaeological Problem**
>
> While excavating an archaeological site, four separate structures are uncovered that seem to be related to each other. The archaeologists suspect that these structures are part of a system of structures that form a circle. (See Figure 3.3a.)
>
> What could be done to check this hypothesis?
>
> If the hypothesis seems to be confirmed, how could they determine the centre and radius of this circle in order to target further excavations? In how many different ways could this be done?
>
> *Figure 3.3a*
>
> *Comment*
> This is a situation to which you can bring your knowledge about circles. Imagine that the archaeologists' hypothesis is true. Draw a simplified diagram of the situation and add in some properties you know.

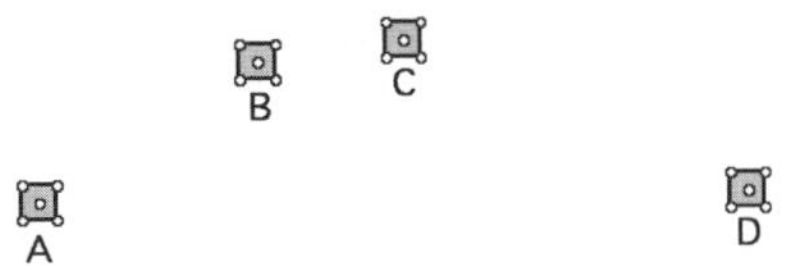

Using the conjecture, a simplified diagram would be a circle with four points A, B, C and D on the circumference. Chords and angles could then be drawn and known relationships checked by measurement at the site. In fact, there are many possibilities: many different chords can be drawn. Very small angles might be hard to measure accurately, and there is imprecision in the choice of points to represent the structures to take into account.

Once you have confirmed that the four points do more or less lie on a circle, there is the problem of finding its centre. By imagining the centre already found, you might think to relate it to various chords. For example, there are infinitely many circles passing through *two* given points. All you can determine in that case is a line that must contain the centre. But with more than one chord available you can determine a unique centre.

In reality, this problem has other archaeological links. The artefacts found on sites often include pottery and broken pieces of plates and saucers. Mathematically, if you have a broken piece of a plate with an edge that might be circular, you can reason very similarly to check for circularity. By finding the radius you can build a template against which to test other shards of pottery.

Geometrical thinking moves from the archaeological context to pure geometry and concludes that the centre of any circle must lie on the perpendicular bisector of each and every one of its chords. This gives a method for constructing the centre given two chords (which only requires three points). Of course the more like a diameter the chords, the more accurate will be the physical construction. Now start with a triangle.

Task 3.3.2 Perpendicular Bisectors of a Triangle

Imagine (or draw) a triangle ABC and treat the three sides of the triangle as three chords of a circle. Construct the perpendicular bisectors of the sides AB and BC and mark the point of intersection D. What happens to D as you drag A, B or C?

Comment

Dragging in dynamic geometry software often reveals more than expected, which is why it is really worth while imagining and conjecturing first, before using software. If you think of the sides of a triangle as the chords of a yet-to-be-constructed circle, then it follows that the three perpendicular bisectors must coincide at the centre and that there is a unique circle through the three vertices of any triangle. Note the scope of generality, of what can change.

Now consider quadrilaterals. Do they have circumcircles? Reasoning might proceed as follows. What would it mean to be a circumcircle of a quadrilateral ABCD? Such a circle must pass through all four points. Now focus on what is known about circles and just three points: there is a unique circle passing through any three of the points, say A, B and C. Clearly, in general, it is very unlikely that this circle will also pass through the point D. Note that this is not the same kind of unlikelihood as expressed by the pupils learning Pythagoras' theorem in the delightful book about a mathematics teacher by H.F. Ellis (1981: 10–11):

> 'Is that a likely thing to happen?' Mason asked. … 'I mean is a right-angled triangle likely to have a square on its hypotenuse? … I mean in real life' …
> 'I see what Mason means, sir,' said Hillman. 'I mean it would be a pretty good fluke if a triangle had squares on all its three sides at once, wouldn't it, sir?'

If the fourth point does lie on the circle, there must be some extra relationships to be discovered. Going back to the relationships discovered in section 3.2, you know that a quadrilateral must satisfy certain constraints in order to be cyclic (for its vertices to lie on a circle). However, once those constraints are satisfied, then the method for constructing its circumcircle will be exactly the same as for a triangle.

Another way to investigate the conditions for a quadrilateral to be cyclic is to construct the circumcircle of three of the vertices, and the perpendicular bisector of another side of the quadrilateral, and move the fourth point around until there are some coincidences. Measuring the angles of the quadrilateral is instructive.

Having considered problems of constructing a circle *around* triangles and quadrilaterals, you might think to try to construct a circle *inside* a polygon. That is, the circle must be tangent to each side of the polygon. Again, for most polygons (quadrilaterals and above) this will only occur if certain further conditions are satisfied. For example, perceptually it is quite obvious that it is impossible to construct a circle passing through all four vertices of the rhombus in Figure 3.3b, and it is equally obvious that it is not possible to construct a circle touching each side of the trapezium in the same figure.

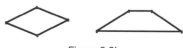

Figure 3.3b

Task 3.3.3 A Circle Inside a Triangle

Imagine or draw a triangle. How might you construct a circle inside the triangle that is tangent to each side of the triangle (the *inscribed* circle)?

Comment

Again, the most useful strategy is probably to draw a sketch of the required diagram and think about the properties it must have. It often helps to imagine that the construction has been achieved, and to look for necessary relationships that suggest a construction. In dynamic geometry software, you can sometimes even reverse the construction: start with a circle and construct three tangents at three different points on the circumference, producing a triangle whose sides are touching the circle.

In constructing a circle the hardest part is to construct the centre. Once you have this, it is relatively easy to construct the circle. Having drawn a sketch, you need to discern some key features that will lead to a construction. For example, if you focus on the fact that the sides of the triangle are tangents to the circle, you might realise that this means that each vertex of the triangle is subtending two tangents to the circle. Task 3.2.6 is useful here. In showing that the two tangents are equal, you probably drew in the line joining the vertex to the centre, and noticed something about the angles of the two triangles. These relationships might suggest how to draw the line from a vertex of the triangle through the centre of the circle.

In the comment above, it was suggested that you could represent the required diagram by constructing three tangents. Will *any* three points on the circumference suffice? Will they always produce a triangle? What are the conditions that do produce a triangle?

The arguments you have used in order to find a construction for the inscribed circle, suggest that the bisectors of the angles of a triangle always coincide in the same point, the *in-centre*. This is because the centre of the inscribed circle must lie on each bisector of the pairs of tangents from the three vertices, and this can be confirmed or evidenced by dragging in dynamic geometry software.

The result of this reasoning is that for any triangle there is a unique circumcircle and a unique *in-circle*. You might now wonder which quadrilaterals have an in-circle. The condition for such quadrilaterals (called tangential quadrilaterals) is that the sums of pairs of opposite side lengths are equal. Conditions are often two-way: if a quadrilateral has a circumcircle then pairs of opposite angles have the same sum (which must be 180°), but also, if both pairs of opposite angles have the same sum, then the quadrilateral must be cyclic. Similarly, if a quadrilateral has an in-circle then the sums of pairs of opposite side lengths must be equal, but also, if the sums of pairs of opposite side lengths are equal, the quadrilateral has an in-circle.

Reflection 3.3

Look back over the reasoning you have done and have followed, and identify what the invariance was, what was allowed to change and in what way. Try to locate shifts in how to focus attention during the reasoning, or how following the reasoning required you to shift the focus of your attention to discerning sub-figures, recognising relationships, imagining elements that were not yet actually present, perceiving properties that sub-figures might have, and then how the reasoning proceeded on the basis of known or defining properties.

3.4 CONSTRUCTION AND DESIGN

Design is a very important and pervading feature of daily life. Almost everything human-made has a 'designer' behind it. Some design is purely decorative, for example, designs in carpet or curtain patterns. Some is almost entirely functional, as with design of machine parts. Other designs are a mixture of both, for example, furniture design certainly needs to be functional but its aesthetic appeal is also of importance. When a design is for purely decorative purposes it may be produced through a good deal of experimentation because perhaps no fixed image or requirement is in the mind to start with. But even then, one often needs something as a basis with which to experiment. For example, the seven circles activity in Chapter 1 (Task 1.2.3) could well form a basis from which one can isolate interesting elements. The three designs in Figure 3.4a are all extracted from the seven circles diagram.

Figure 3.4a

If you want to design something for a functional purpose, you are usually confronted with constraints that have to be met. This involves *constructing* a geometrical figure that meets those constraints, rather than selecting elements from an already existing figure. In fact, some of the tasks you have already done could be thought of as design tasks. Construction tasks feature in both parts of this section.

Construction with and of Circles

The final tasks in the previous section were concerned with the problem of inscribing a circle within a given polygon. Now consider the same idea between the polygons themselves. For example, in Chapter 1 you looked at the problem of constructing a rectangle inside a square or another rectangle. What about constructing an equilateral triangle inside another triangle?

Task 3.4.1 Inscribed Equilateral Triangles

How many equilateral triangles can you inscribe in a given triangle (one vertex on each of the sides of the original triangle)?

Comment
A natural thing to do is to start with a special case: an equilateral triangle. This might trigger recognition of having seen an equilateral triangle inscribed in previous tasks, formed by joining the mid-points of the sides. Sketch as if you had found one, and then try to see what relationships are forced. Alternatively, use Polya's idea of letting go of one constraint (allow one vertex not to be on the appropriate side of the triangle) and then see what path it traces as you change the size and orientation of the equilateral triangle. Finding triangles is just the beginning. What relationships are required, and how then might they be constructed directly?

The interactive file '**3c Two Equilateral Triangles**' shows one solution. Since there are many solutions, it is natural to ask what characterises the inscribed triangle with the largest and smallest areas.

A defining property of an equilateral triangle is that it has two angles of 60°. An attractive alternative solution uses this property, that is, trying to construct an inscribed triangle with two angles of 60°. It is not hard to construct three angles of 60° at any point P on the base BC, as shown in Figure 3.4b.

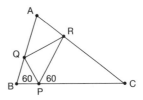

Figure 3.4b

Although triangle PQR is *not* equilateral, it does have one angle of 60° opposite side QR. This angle could be described as being *subtended* by QR, which might well bring to mind circles (see Figure 3.4c). One of the fundamental invariant properties of circles is that a chord subtends equal angles at the circumference. Hence, by drawing a circle through points P, Q and R you can effectively 'shift' the 60° angle to a new position in the figure. So what? Why should the triangle QRS be equilateral? Focusing on the known angle ∠QPS and discerning the fact that the PQRS is a cyclic quadrilateral shows that another angle, ∠QRS, must also be 60°.

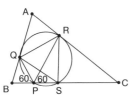

Figure 3.4c

Having seen how circles can be used in constructions, you will now see how dynamic geometry software can make exploration of a more challenging construction more accessible.

Task 3.4.2 Circle Tangent to a Line and a Circle

Imagine constructing a circle tangent to a given line at a given point. Imagine constructing a second circle, having the same radius as the first, tangent to both the line and the first circle.

Given a circle tangent to a given line, imagine constructing a second circle tangent to the first circle and tangent to the given line at a specified point.

Given a circle tangent to a given line, imagine constructing a second circle tangent to the given line and tangent to the first circle at a specified point.

Comment
You will probably find that when the two circles have equal radius, the construction is not too difficult. Notice how you need to imagine or draw a sketch in order to appreciate what is being asked. The second and third cases require more powerful constructions so that any radius may be chosen for the second circle. Two circles of equal radius will then be a special case of the more general construction.

Having found an approximate method, attention can turn to a precise construction. In either case, you can determine a construction by looking at the properties of the figure when the two conditions are met, as shown in Figure 3.4d. Notice that the fact that B is a point of mutual tangency suggests drawing in the common tangent at B, producing the point C.

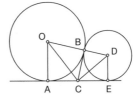

Figure 3.4d

From all the circle relationships you have discovered earlier, you can probably list many relationships in this diagram. For example, B lies on the line OD; BC is tangent to both circles; CA = CB = CE (equal tangents from a point); the angles at O and D are bisected by OC and DC respectively; DE is perpendicular to CE. Now you can choose which of these properties will help you to construct the figure from a particular starting point.

Suppose you want to construct the second circle to touch the first circle at a given point, like B in Figure 3.4d. The first property listed above says that the centre of the second circle must lie on a line through OB. Focus on B, and look for other elements that could be constructed, knowing B. A few constructions lead to locating D. Similarly, suppose you want to construct the second circle to touch a given point on the line, like E in the diagram. The last property listed above tells you that the centre of the circle must lie on a line from E, perpendicular to the given line. Focusing on E does not yield much else of help, unless you recall the relation between AC and AE. Focusing on C can then suggest a construction to find B and finally D.

The interactive file '**3d Line and Touching Circles**' provides two constructions along these lines. When you drag E or B, the angle ∠OCD is suggestively invariant, which might direct attention to the four angles at C, how they are equal in pairs and all sum to 180°, showing that the angle is invariant.

This approach to investigation using dynamic geometry software illustrates two very important aspects of the problem. First, you can *choose* to construct the second circle starting from either position; that is, a point on the circumference of the first circle, or a point on the line. Whichever you choose will become the *independent* point of your construction. This means that the radius of the second circle will vary according to how you vary this point. Second, the choice of starting point will also determine how the construction proceeds. In other words, you will have two different constructions depending on your choice.

Two Historical Vignettes

This section concludes with two historical vignettes that provide interesting examples concerning circles. The first is in fact a theorem; the second is an application of circle properties. There are, of course, a vast number of theorems in geometry, some of which are considered to be important enough to be included in most curricula around the world. Probably the first to spring to your mind is Pythagoras' theorem. Unfortunately, it is also the case that some theorems could almost be described as 'forgotten' theorems, in that they rarely appear in secondary school geometry texts. One of these is a powerful theorem derived by Claudius Ptolemy in about AD150. It appears in an astronomical treatise, which later became known as the Almagest, meaning 'the greatest'! In fact, Ptolemy is best known for his work in astronomy and trigonometry, and some well-known trigonometric relationships can be derived from his geometric theorem.

Using the circle property of equal angles subtended by equal chords, it is not difficult to derive some relationships concerning lengths. Consider the diagram in Figure 3.4e.

Figure 3.4e

Triangles AED and BEC are similar, and so are triangles AEB and DEC. Using the first of these pairs you can deduce that AE/ED = BE/EC and therefore (AE)×(EC) = (BE)×(ED). You would get the same final result using the other pair of similar triangles, and it is known as the Intersecting Chord Theorem. It remains true even when the chords intersect outside the circle. However, Ptolemy's genius enabled him to derive a much more important relationship between the lengths of the cyclic quadrilateral ABCD.

Task 3.4.3 Investigating Ptolemy's Theorem

Draw a quadrilateral ABCD inscribed in a circle. Measure the lengths of the sides and the diagonals of ABCD. Calculate the product of the diagonals and the product of each pair of opposite sides, that is, (AC) × (BD), (AB) × (CD) and (BC) × (AD). What relationship between the products might be conjectured?

Comment
You will probably be able to spot an apparently invariant relationship quite quickly. However, proving it is not so easy, which makes you wonder how Ptolemy spotted it! It cannot be derived using only the similar triangles identified earlier: some further insight is required.

The proof of the theorem depends on deliberately *introducing* another similar triangle into the diagram. Look at the interactive file '**3e Ptolemy**' to see how the theorem is derived. As well as deriving trigonometric relationships from the theorem, a simpler, very striking relationship can be quickly established by trying it on a special case. Consider the special case of a rectangle whose sides are of length a and b. Being a rectangle it is necessarily cyclic. Now apply Ptolemy's theorem: the sum of the products of opposite sides is the product of the diagonals. Labelling the lengths of the diagonals as c produces the familiar theorem that lies at the intersection of geometry and algebra: Pythagoras! This is a nice example of how different areas of mathematics link together and how one can become a confirming instance of another.

The second historical vignette looks at how knowledge of circle properties can help solve a problem first posed in 1471 by the German astronomer and mathematician known as Regiomontanus. As quoted by Maor (1998, p.46), the problem was expressed in the following way: 'At what point on the ground does a perpendicularly suspended rod appear largest?' Put in this way, the problem sounds rather abstract, but it can be put into a real-life context by imagining a statue on a tall column and again asking what is the best position from which to view it so that it appears largest. Gallery-goers can think of a painting high on a wall; rugby players can think of placing the ball for a conversion. In Task 3.4.4, the idea of 'appears largest' is re-expressed in a simpler mathematical form.

Task 3.4.4 Regiomontanus' Problem

The diagram in Figure 3.4f shows a statue on a column.

Find the position at which the statue subtends the greatest angle at the person's eye.

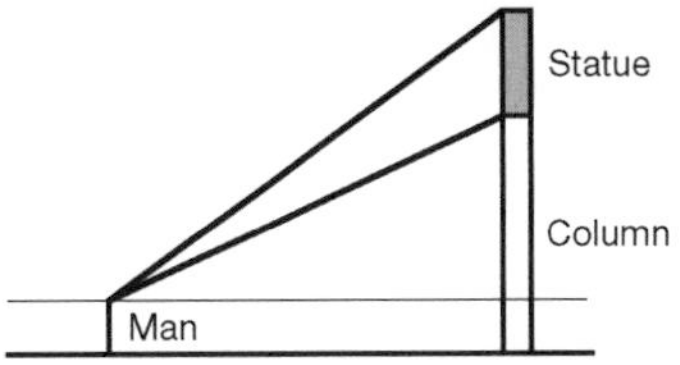

Figure 3.4f

Comment
No lengths are given, and there are no obvious circles around. You could use some specific values. For example, the height of the column might be 6 m and the height of the statue 2 m. The person's eye might be 1.6 m above the ground. The aim is to think geometrically rather than arithmetically.

Think of the statue as the chord of a circle and the angle subtended at the viewing point as an angle at the circumference. What circle properties come to mind triggered by that language?

This problem can be solved algebraically, of course, but this was not available to Regiomontanus! Moreover, it is often the case that a purely geometric solution is simpler and more elegant. As you have seen before, it is easy to construct a circle with the statue as chord but what will ensure that the angle subtended at the viewpoint is a maximum? Imagine the viewpoint coming closer and going farther away, and how the subtended angle changes (see Figure 3.4g). Consider the diagram on the right, which shows a circle with the statue as chord and passing through the viewpoint. Why does that viewpoint *not* give the maximum angle?

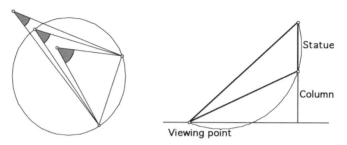

Figure 3.4g

From these observations you can conclude that the maximum subtended angle will be when the horizontal line through the viewing point is a tangent to the circle with the statue as chord. So the problem now becomes finding a method to construct this circle. More specifically, how can the position of the centre be determined? The circle must pass through the top and bottom of the statue. Consider the invariant properties of the situation. What do you know about a chord in relation to the centre of a circle? From this you can construct a line that the centre must lie on. Now focus on the viewing point (the person's eye). Because the line through the viewing point must be a tangent to the circle, the height of the centre above this line must be the radius of the circle. So now the radius is known and hence the centre can be found using one of the points, say the top, of the statue. A dynamic geometry software construction helps in verifying that no other position on the ground has a larger angle subtended by the statue. (See the interactive file '**3f Regiomontanus**'.)

Reflection 3.4

What is the same and what is different about the reasoning involved in Tasks 3.4.1 and 3.4.2? And in Tasks 3.4.3 and 3.4.4?

3.5 PEDAGOGIC PERSPECTIVES

Because of the many invariant relationships connected with circles, they can be useful when reasoning, and they are involved in constructions that have nothing ostensibly to do with circles.

Inner and Outer Tasks

The tasks offered in these first three chapters have revealed a large number of geo-metrical facts or *theorems*. The word theorem is of Greek origin and means, literally, a 'seeing'. That has been their overt or *outer* purpose. But the tasks were chosen in order to afford you experience of different aspects of geometrical thinking and, in this chapter particular, geometrical reasoning. This was part of the *inner* purpose of the tasks, which could not be stated explicitly because then you would have been attend-ing to that and not gaining the experience. By engaging in the explicit outer task, you will have had opportunities to discern details of figures, to recognise relationships, to perceive properties (independent of the particular figure) and to develop your reason-ing on the basis of properties. You may not always have been aware that this was going on, but over a period of time working through this book you will be building layer upon layer of experience of subtle but important aspects of geometrical thinking.

Two Pedagogic Strategies: Say What You See and Same and Different

When pondering a figure, you will have noticed that a great deal rides on noticing elements and relationships that can be developed into a chain of reasoning. One of the best ways to help yourself, or a group of learners, to discern relevant and useful elements, sub-figures and relationships is literally to 'say what you see'. Each person can say something, and each element pointed out will direct attention. You soon dis-cover that what is salient for one person is overlooked or ignored by another. Developing geometric thinking depends on learning to discern and to recognise geo-metric relationships that are fruitful, and this is best developed through participating in a group and hearing what others have to say.

You will have noticed the occasional use of the question 'what is the same, and what is different about … ?'. This turns out to be an extremely fruitful type of ques-tion when there is more than one learner pondering two or more figures or sub-figures. What is the same and what different about two triangles can lead to seeing that they are congruent or similar; about two angles can lead to seeing that they are both subtended from the same chord of a circle, and so on.

The Role of Imagination

Powerful as dynamic geometry software is for revealing relationships, the most power-ful tool is your own imagination. Whenever possible it is wise to try to imagine a figure before sketching it, to sketch it before using dynamic geometry software, and when using dynamic geometry software, to make a conjecture as to what will happen before actually doing it (dragging, constructing a new element), thus creating an expectation. Expectation is a product of imagining, and without an expectation there is unlikely to be any surprise, yet geometry is a domain full of surprising relationships.

Several times in tasks calling for a construction, it has been useful to imagine the construction *as if* it were completed. This introduces extra elements into a figure, which in turn can help you to recognise relationships, perceive these as properties and so, support your reasoning on the basis of properties.

Unexpected Invariants

In Task 3.2.5, it was shown that angles subtended on the same side of a chord in a circle must be equal. With care it is possible to define what you mean by 'angle subtended by a chord in a circle' so as to remove the need to specify 'on the same side'. It is also the case that if you start with the property 'all angles subtended (on the same side) of a segment are equal', then the points satisfying that will lie on a circle with the segment as chord.

Many of the tasks in this chapter have attempted to highlight how the invariant relationships (perceived as properties independent of the particular figure) play an important role in the reasoning process when solving problems or designing constructions. It quite often happens that textbook problems, and students' attempted solutions, will unexpectedly raise interesting issues that when probed, reveal invariances. The following two tasks illustrate this. The first is ostensibly a very simple problem concerning circles.

Task 3.5.1 Three Equal Chords

In Figure 3.5a, AD is a diameter and AB = BC = CD. Show that BC is parallel to AD.

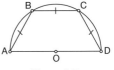

Figure 3.5a

Comment
You may not be surprised that most people approach this problem by drawing in the radii OB and OC. (You might like to give it to some learners to see how they tackle it.) If this is what you did, try finding another way to show BC is parallel to AD, and compare the information and concepts you have used in both methods.

The radii approach mentioned in the comment above divides the diagram into three congruent isosceles triangles and hence $\angle AOB = \angle BOC = \angle COD = 180°/3$. This forces the triangles to be equilateral, so in particular $\angle AOB = \angle OBC$. Since these are alternate angles, BC must be parallel to AD. Compare this to the solution given by one student (Figure 3.5b).

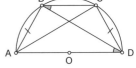

Figure 3.5b

In Figure 3.5b the line BD is drawn. $\angle ADB = \angle DBC$ because they are subtended by equal chords (AB and CD). But since they are also alternate angles, BC must be parallel to AD.

It is certainly very different to the radii approach and is striking in its simplicity and elegance. Apart from the final step concerning alternate angles, the concepts used are quite different too. In fact, it just uses the invariant property of equal angles subtended by equal chords in a circle. But what about the information used? Again, there is a striking difference here. The fact that AB = CD was crucial, but the fact that BC is also equal to these sides was ignored. This means that it must be redundant information as far as the parallelism of BC and AD is concerned. Not only that, but also the fact that AD is a diameter was ignored, so that too must be redundant. That is, the original diagram can now be seen as a special case of a more general invariance: if two equal chords AB and CD are drawn *anywhere* in a circle (with A, B, C, D in order round the circle) and the points BC and AD are joined, then BC is parallel to AD (indeed, AC = BD as well!). Draw this in dynamic geometry software and drag the independent points around the circumference to confirm it. Have you also noticed that this has a similar taste to Task 3.2.4? Here what is given and what is to be proved have been interchanged. This makes Task 3.5.1 the *converse* of Task 3.2.4, and vice versa. What happens if AB and CD cross each other?

The final task also involves an assumption on the part of a learner but this time it leads to seeing how invariance can be used as a way to solve a problem.

Task 3.5.2 Unwarranted Assumption?

The diagram on the left in Figure 3.5c shows a geometric problem given to a class of secondary students. In this problem the value of *x* (that is, ∠BCD) has to be found. There are a number of different ways to solve this. See how many you can find.

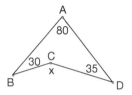 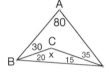

Figure 3.5c

The diagram on the right shows one student's solution. He claimed the triangle ABD was isosceles. From this, he found its base angles and then he calculated the base angles of triangle BCD. Angle *x* was found by calculation. The final answer is the correct solution. Explain this.

Comment
The original diagram does not show AB = AD. When asked by the teacher, the student said that the triangle looked isosceles. His method appears to be unjustified and teachers frequently warn students, quite rightly, not to make unwarranted assumptions. Is this example just a coincidence? Does it matter what angles CBD and CDB are as long as they add up to 35°?

Coincidences have been mentioned in the previous chapters and it was noted that mathematicians are deeply suspicious of them. However, if it is not a coincidence there must be some underlying explanation. You could try the same reasoning with other numbers (specialising is often a useful strategy). You could also try working with the angle relationships arising from triangles when you insert the side BD into the figure but without making the assumptions made by the learner. You will find that what matters is the sum of angles CBD and CDB, not their values.

The class in question decided to 'discern' the isosceles information by inserting it into the diagram. In Figure 3.5d, AE is constructed equal to AB. From the point E a line is drawn parallel to DC, to intersect BC produced, at F.

The effect is to shift the angles *x* and 35° to another part of the diagram. The line CD can now be ignored. Joining BE produces exactly the diagram the student had assumed. This is certainly

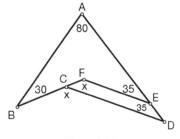

Figure 3.5d

not the most elegant way to solve the problem; perhaps you found that one of your own previous solutions is easier. But it illustrates quite vividly that, by using an invariant property of a situation, you can sometimes convert a difficult problem to a simpler one. This example is reported by Lopez-Real (2003) and perhaps the following quote of the teacher's reaction, after she discussed this with some fellow teachers, makes a fitting conclusion to this chapter:

> I took it back to my class and they were all so excited to see why it could work. It made us look at the diagrams in some other problems to see what we could change and what would stay the same.

Reflection 3.5

What difference did the availability of the interactive files make to you in this chapter?

4 Visualising and Representing

The central question in this chapter is what makes a presentation or re-presentation of spatial relationships useful for thinking and for communicating effectively with others. For example, if you wanted to inform someone about the structure and setting of the London Eye landmark, and you could only show them one picture, you would probably choose the third of the ones shown in the photographs (Figure 4.0). The middle picture yields less information about the Eye than the other two, and the first picture gives less information about the location than the third. Each is a two-dimensional projection of a three-dimensional object, yet children quickly learn to interpret such images three dimensionally.

Figure 4.0

In this chapter, attention is focused on different ways of presenting three-dimensional objects in two dimensions. The first section includes different methods of projection in regular use. The second is about plans, elevations and slices used by engineers and others. The third section is about the geometric thinking which underpins western art. The fourth section has a focus on paths traced out by movement of constrained parts. The chapter ends with a pedagogic review.

4.1 VISUALISING THREE DIMENSIONS USING ONLY TWO

Instructions for assembling furniture, for laying cork tiles and other flooring, and for assembling Lego models all make use of diagrams that require interpretation in order to make sense of them. To gain some appreciation of what might be involved, this section has some tasks that invite you to imagine three-dimensional objects formed by rotating a two-dimensional object about an axis, and to consider what is involved in communicating through diagrams.

Task 4.1.1a Revolutionary Solids

Figure 4.1a shows a rectangle (think of a piece of card) and various axes. For each axis, imagine the rectangle being revolved (rotated) through 360° about that axis. Imagine what the solid shape swept out would look like from the outside. What makes the visualisation easier or harder?

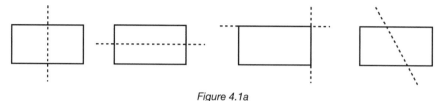

Figure 4.1a

Comment

You may have found the vertical axis easier than the horizontal one, and found it more difficult to ignore one axis when two were presented on the same diagram. Most people find the sloping axis most difficult, especially because it is hard to tell what is visible when the rotation is finished and what is not.

Having imagined a solid shape, how might it be recorded? There is a spectrum of responses to this question, from giving people instructions like those in the task, through using a label such as *cylinder*, to sketching it from an angle so as to depict its principal features.

This task is a glimpse into a vast domain of tasks in which shapes are revolved around axes: triangles, squares, other polygons and even circles can be rotated about axes and the swept-out shape can be imagined and sketched. Furthermore, you can then reverse the process as in the next task.

Task 4.1.1b Seeing Solids as Revolutionary

What is the same about all the solids created in the previous task, and what is different about them?

Which solid shapes that you encounter daily could be formed by rotating a two-dimensional shape around an axis, and which could not?

Comment

All the rotated objects have an axis of symmetry, but most importantly, they all have circular cross-sections perpendicular to that axis. Bottle corks have the shape of a frustrum (a cone with its top cone sliced off parallel to the base), as if produced by rotating a symmetric trapezium around its axis of symmetry; funnels, cylinders and hemispheres are some more of the shapes you can find (Figure 4.1b).

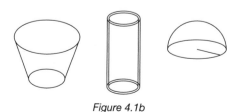

Figure 4.1b

The next few tasks develop some standard conventions for depicting three-dimensional objects in two-dimensions. For example, word-processing packages often allow you to draw basic two-dimensional shapes and convert them into images of associated three-dimensional solids. Starting with the square in Figure 4.1c, it is possible, for example in the word-processing software used to write this chapter, to create an image that gives the impression of a cube.

The options selected to draw the two cubes in Figure 4.1c are based on a method known as 'oblique projection'. The shaded diagram obscures details and this makes interpretation ambiguous. The 'wire-frame' has its weaknesses as well. You will explore some of these weaknesses in the next task.

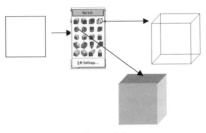

Figure 4.1c

Task 4.1.2a Many Cubes

What is the same, and what different about the depictions of a cube in Figure 4.1d?

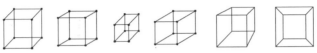

Figure 4.1d

Comment

In the first four diagrams, parallelism is preserved. Some preserve equal lengths. Some look 'more like' a cube than others; this is because you are more used to 'seeing' some more than others. Did you notice that in all of the depictions there is ambiguity as to which 'face' is the front?

The first four diagrams of Figure 4.1d are called *oblique projections*: they project a front face to a back face and preserve parallel lines (edges parallel in the solid are shown parallel in the diagram). By focusing on one face 'at the back' and ignoring a parallel face at the front, you can make the solid shape 'flip' so that it looks inside out. Prisms are solids for which all the sections parallel to one face are identical and are usually depicted using oblique projection.

Task 4.1.2b Flipping Figures

What do you need to do in order to make the diagrams in Figure 4.1e 'flip'?

Figure 4.1e

Comment

Dynamic geometry software enables you to experiment with the proportions and orientation of the front and back faces in a diagram, to see which are easier to 'flip' intentionally. See, for example, the interactive files '**4a Triangular Prism**' and '**4b Oblique Cubes**'. Even a shaded drawing can be made to flip with some mental effort. By focusing attention, it is possible to force edge AB of the shaded cube to come closer, and consequently make CD recede.

If you did not manage to make the shaded cube 'flip' inside out, just keep practising!

Isometric projection is another method commonly used to represent three-dimensional objects on paper. Design and technology lessons tend to use an isometric grid that has vertical lines and two sets of parallel lines inclined at 30° to the horizontal (resulting in a tessellation of equilateral triangles) whereas mathematics lessons tend to use isometric 'dotty paper'. (See Figure 4.1f.)

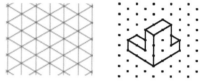

Figure 4.1f

You may find it useful to use some children's building blocks, multilink cubes, dice or sugar cubes for the next task.

Task 4.1.3 Constructing and Deconstructing Solids

How many different objects can be made from cubes that, when sketched from the correct angle, would look like the isometric drawing in Figure 4.1f?

Sketch an object (as if) on dotty paper that uses only five cubes, but which is unambiguous from one direction and ambiguous from another. There is a file containing **a worksheet of isometric dotty paper** on the CD-ROM.

Comment

Some learners find dotty paper difficult to use, perhaps because they are not used to interpreting such diagrams. Many people find it difficult to imagine parts that are hidden from view.

The best way to test whether you understand something is to write instructions to someone else or, better yet, to a computer, which will not 'interpret' what you mean from what you say.

Task 4.1.4 Oblique Instructions

Write brief instructions that someone else could use to draw a prism using oblique (or isometric) projection.

Comment

Did you think to indicate what remains invariant, and what can change? Did you find it hard to be clear? Did you persevere and try to write down instructions?

The intention behind the tasks in this section is to provide experience that involves manipulating images and dynamic geometry software figures in order to get a sense of problems with depicting three dimensions in two dimensions. Task 4.1.4 ends the sequence with an invitation to try to bring that sense to articulation. It usually takes multiple experiences before sufficient clarity is achieved to enable learners to give clear instructions.

In the next section, techniques for displaying three dimensional objects will be developed that produce unambiguous drawings.

Reflection 4.1

Have you noticed learners having difficulties similar to the ones you experienced? What might you do to gain more experience for yourself, and how might that experience help you be sensitive to learners' needs?

4.2 PLANS, ELEVATIONS AND SLICES

Engineers and technical draughts-persons overcome potential ambiguity by providing three different projections or view points (plans and elevations) and, for good measure, they also put in hidden lines corresponding to edges that may not be visible. These are particularly necessary if the object is actually hollow rather than solid! Before considering slices of three-dimensional objects as two-dimensional shapes, a solid shape is used as a case study in how and why technical drawings are used. You may find it useful to have such a solid to hand when working on this task – you may be able to build one using multilink cubes or a construction toy such as Lego.

Task 4.2.1 Looking in Different Directions

Make or imagine the solid shape shown in Figure 4.2a, and assume that it is made up of cuboids with no holes or parts sticking out round the back. From which direction would you see the projection in Figure 4.2b? Sketch the projections you would see looking from each of the other marked directions, including in your sketch any edges that are visible from that direction.

Figure 4.2a   Figure 4.2b

For each of the solids shown below, work out which projection would also be the projection in Figure 4.2b.

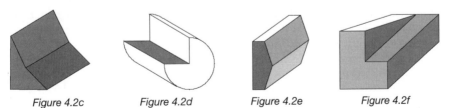

Figure 4.2c   Figure 4.2d   Figure 4.2e   Figure 4.2f

Comment
From direction C, you would see what is in Figure 4.2b but you would see the same thing for the solids in Figures 4.2c, d and e as well. The solids in Figures 4.2a and 4.2c, d, e and f would all be distinguished by a projection from direction A. All five of the solids would have identical projections from one direction. What is needed is a collection of projections in order to remove ambiguity.

The sketches you have drawn are known as elevations. It is common practice to choose one of these as the 'front' elevation (usually one that shows the most significant detail of the object) and the others are then known as the side (or end) elevations. The view from above is known as the plan. In the case of the L-shaped block (Figure 4.2a), viewing from the direction of arrow A gives a front elevation, which gives the cross-section of the prism. The views in the direction of arrows B and C give end (or side) elevations, and viewing in the direction of D gives the plan.

Learners cannot be expected to master visualising and sketching projections from different directions with just a few experiences. Most will need to develop a rich fund of experience over a considerable period of time, gradually learning to express in words the sense that they are making of their experience. Furthermore, by using undoing tasks, in which projections are given and an object has to be made to match those projections, learners gain experience both seeing three dimensions in two dimensions, and using two dimensions to see three-dimensionally.

For technical drawing, it is essential to include three projections of an object in one diagram, and there are two standard ways of doing this: exploded diagrams and orthographic projections.

Exploded diagrams are achieved by thinking of the object as sitting inside a box with transparent walls, and imagining that views of the object are projected orthographically (at right angles) onto the wall that lies between the viewer and the object.

Orthographic projections go further, by imagining the box unfolded to reveal the net with all of the views of the object visible. As Figure 4.2g shows, this process results in the plan placed above the front elevation and an end or side view placed to the right of the front elevation.

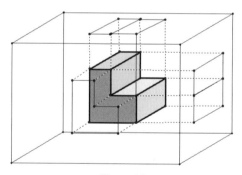

Figure 4.2g

Computer software has made the process of drawing objects much easier, since objects can be manipulated as if they were being rotated in space, so that they can be viewed from all directions. Previously, technicians had to master techniques for drawing the three projections. An example of two of the steps in the development of the L-shaped solid is shown in Figure 4.2h.

The extra lines drawn in enable the technician to transfer measurements

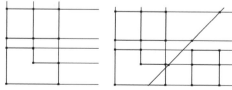

Figure 4.2h

accurately from one part of the drawing to another. A useful way to check understanding of how this process works is to try an example for yourself (for example, modifying Figure 4.2h to depict the solid in Figure 4.2c), to try 'undoing' or reversing someone else's drawing, or to ask yourself questions about what can be changed and what must remain invariant in the technique (for example, must the angle of the diagonal line in the last drawing be 45°?).

In order to improve communication, it is usual for engineering drawings to include dotted lines to indicate edges that are hidden when looking at the object from that particular direction. One topic for further exploration is, if hidden lines are shown, does this remove the need for three projections?

The ideas developed so far can now be used to explore slices of solids such as a cube.

Task 4.2.2 Slicing a Cube

Imagine slicing through a cube with a single plane. Three of the many possibilities are shown in Figure 4.2i, giving cut faces of two rectangles, and a triangle. What would the missing portions of the cubes look like? How many different polygons can be the cut face of a cube? (You must decide what counts as different; a good challenge is to characterise the ratios of the edges of rectangles, and other polygons.)

Figure 4.2i

Comment

You may wish to reach for an interactive file here, but it is more useful to try to imagine them for yourself first, and then to test your conjectures by cutting through solid shapes. A cube made of plasticine or cheese might be helpful.

As well as imagining different slices, you can decide on a polygon and then try to see if such a slice is possible. Did you find an equilateral triangle, an irregular quadrilateral, a pentagon and a hexagon? Why is a heptagon not possible?

Were you content with a single dotty paper type sketch or did you reach for the plasticine? Were you motivated to draw three orthographic projections in order to strengthen your grasp of that technique?

As a variation on the task, you could also find two different ways to slice a solid cube so that two or three elevations stay the same.

Cube slicing can be great fun if it is done in a group, because many conjectures will be formulated, and then contradicted or justified. Other aspects of the task that can be varied include slicing another familiar shape (for example, a cuboid or other prism, or a regular tetrahedron), or making two or more slices.

The next task gives you the opportunity to work in the reverse direction, that is, from two dimensions to three.

Task 4.2.3 Rebuilding a Sliced Cube

The projections shown in Figure 4.2j are of a cube that is cut by a single plane face that cuts two edges of the cube in their midpoints, and one edge of the cube at a one-third point. Reconstruct the cube showing where it is sliced.

Figure 4.2j

Comment

It is not as easy as it might seem at first. Where did you start when trying to sketch the cube: with a cube that you then 'sliced', or with the slice itself, or with the part of the solid not touched by the slice? Being aware of choices is very useful because if one approach runs into difficulty, you know you can try another approach.

Did you use dotty paper or plasticine? What properties does the triangular slice have (what kind of triangle is it, what are the ratios of its side lengths)?

One method of constructing an accurate view of the cut triangular face in the previous task is known as an *auxiliary projection*, which is a method of drawing a projection from a useful direction. Step-by-step instructions are given in Figure 4.2k. Notice what you have to do to make sense of the diagram and the words, by moving your attention back and forth.

(i) Produce construction lines orthogonally (at right angles) to the cut face. These ensure that the triangle you construct will have the correct height.
(ii) Drawn an axis perpendicular to the projection lines.
(iii) The length of the base is taken from the appropriate view – in this case the plan.
(iv) The resulting triangle will have the dimensions of the cut face.

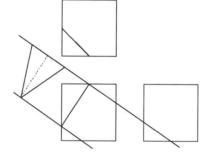

Figure 4.2k

Historically, auxiliary projections were used to study the two-dimensional shapes that arise from sections of a cone. The curves are often referred to as *conics*. It is thought they were first studied systematically by Menaechmus who was a mathematics tutor to Alexander the Great (approx 350 BCE). Imagine slicing through a cone with a single plane. The aim is to find out what two-dimensional shapes can arise. Note that the cone has to be complete: as well as the familiar conical shape there is another conical part upside down above the vertex.

The easiest case is a slice perpendicular to the axis of the cone, parallel to the base. The section will be a circle, and this is shown in two projections in Figure 4.2l. The line AB in the front view shows where the edges of the cutting plane intersect with the sides of the cone; the large circle is the 'base' of the cone, and the small circle is

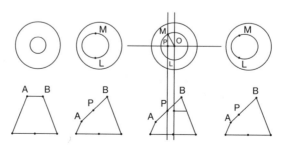

Figure 4.2l

the actual section slice. The circular section will form the top circular face of the resulting solid (and the bottom circular face of the part cut away).

If the slicing plane is at an angle to the vertical axis of the cone, then the cone viewed from the side is as shown at the bottom of the second sketch in Figure 4.2l. The resulting section is an ellipse. However, the plan does not give the actual dimensions of the ellipse, because the ellipse is not parallel to the base. Although the plan shows an ellipse, it is in fact a foreshortened version of the slice. It is possible now to use an auxiliary projection to construct the actual shape of the ellipse.

Consider the point P in the third sketch in Figure 4.2l. The plan shows two points, L and M, on the edge of the ellipse that both correspond to P in the elevation. To determine the plan view of the ellipse you need to know where L and M are, for all positions of P.

For each position of P, points L and M lie on a circle with centre O (in line with the vertex of the cone). The radius of this circle OM has the same radius as the horizontal circle that would be formed by a horizontal plane cutting the plane that passes through P. Although it may take a moment or two to sort this out, it is an excellent opportunity to see how what at first seems difficult and strange can actually be made sense of.

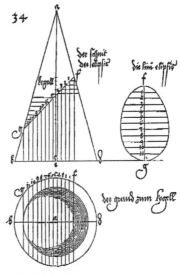

Dynamic geometry software is ideal for doing this sort of geometrical reasoning, ending with the trace (or locus) of the points L and M to see the ellipse: open the interactive files '**4c Sliced Cube**' and '**4d Slicing Cones**'.

Many artists in the sixteenth century did this sort of work in order to master the art of realistic drawing and painting. For example, the German artist and mathematician Dürer suggested the elliptical cross-section can be found in his *Treatise on Measurement* of 1525 (Figure 4.2m). You will see more of Dürer's contribution to mathematics and art in the next section.

Figure 4.2m

Reflection 4.2

More valuable than trying to master the techniques of projections is to observe the various states of confidence and loss of confidence, of smooth flow and struggle that you experienced. How can learners be helped to appreciate that some struggle, some challenge is worthwhile, because of the feeling of satisfaction it brings when finally things slot into place?

4.3 EXPLORING HIDDEN GEOMETRY IN WESTERN ART

The next part of the chapter focuses on how artists learned to reduce a three-dimensional world to a canvas so that it appears real. A technique of drawing in perspective was known to the artists who decorated walls in Pompeii, but was lost as a result of the eruption of Vesuvius. It was then rediscovered and reconstructed in the Renaissance. Artists studied geometry in order to understand how to achieve realism in their paintings.

In a world of digital cameras, everyone is familiar with interpreting the depth in photographs, but it is something that is learned by young children. It is not 'obvious' from birth. The photograph (Figure 4.3a) highlights features that provide depth cues that a young child might misinterpret. Things farther away are smaller, and lines that are parallel in three dimensions are *not* shown parallel in the picture.

Figure 4.3a

You can get further understanding as to how depth perception works by studying William Hogarth's use of the rules of perspective to fool the viewer (see Figure 4.3b). The picture was included in a mathematical treatise of 1754 by Brook Taylor: *Method of Perspective*.

One of the first to write about the rules of perspective from a mathematical viewpoint was Alberti (1404–72), in two treatises, one of which, *De pictura* written in Latin, was addressed to scholars while the other, *Della Pittura*, written in Italian, was perhaps aimed at a more general audience. In the former of these Alberti writes: 'A painting is the intersection of a visual pyramid at a given distance, with a fixed centre and a defined position of light, represented by art with lines and colours of a given

Figure 4.3b

surface.' The illustrations in Figures 4.3c and 4.3d show how the German Dürer (1471–1528) illustrated Alberti's ideas in a number of woodcuts.

Figure 4.3c

In a woodcut produced in 1525 (Figure 4.3c) the lute is shown being drawn in the plane of the wooden frame from the viewpoint of the 'eye' screwed into the wall. Notice how the picture has been swung out of the way so as to mark a point where the string, representing a ray of light from a point on the lute to the 'eye' crosses the frame. This point is therefore projected onto the plane of the picture where the string intersects the plane. In this way every point in the picture of the lute could be found, somewhat tediously!

In the second woodcut (Figure 4.3d), an artist is shown keeping his eye at a set viewpoint, which determines where rays from his subject will intersect a vertical grid. He then transfers this information to the grid on his drawing (without moving his head!).

Figure 4.3d

One of the famous examples used by Alberti to describe his geometrical understanding of perspective is that of pavimento – pictures that include square tiled floors. Alberti's methods were later simplified by Piero della Francesca (1416–92) who was born near Florence. Della Francesca combined his skills in mathematics and art to write a number of treatises about how to achieve depth in paintings and drawings. One of these, *On Perspective* is written very much   as a book about geometry, laid out in the style of Euclid's *Elements* with theorems and proofs.

Piero articulated what Dürer later illustrated:

> First is sight, that is to say the eye; second is the form of the thing seen; third is the distance from the eye to the thing seen; fourth are the lines which leave the boundaries of the object and come to the eye; fifth is the intersection, which comes between the eye and the thing seen, and on which it is intended to record the object.

Getting a tiled floor to look right is no mean feat, because the proportions need to be just right (and the same technique enables objects to be drawn at heights appropriate to their distance away in the picture). Drawing tiled floors became a mark of an accomplished and skilled artist, until techniques were developed for doing it geometrically. This section ends with a description of how it can be done and an interactive file '**4e Vanishing Point**' so that you can experiment and appreciate the sophisticated reasoning involved.

The technique for drawing in perspective makes use of two observations:

Parallel lines are depicted as lines passing through a common point called the vanishing point for that family.
Vanishing points for different families of parallel lines all lie on a single line in the picture called the *horizon*.

Figure 4.3e shows a chessboard–style grid.

Figure 4.3e

The interactive file '**4e Vanishing Point**' makes it possible to explore how a grid might appear if viewed from different viewpoints.

Using a vanishing point is a good start, but there remains the problem of spacing. How are the sizes of the tiles chosen so as to produce an appropriate sense of foreshortening? Some step-by-step advice about how to construct a tiled pavement is given in Figure 4.3f overleaf. Apart from the edges of the tiles that form two families of parallel lines, there is a third family: the main diagonals (and, of course, other diagonal families as well). Each family will have its own vanishing point, and the relative positions of these will determine the foreshortening effect.

Altering the positions of V and T in the interactive file '**4e Vanishing Point**' affords insight into various effects that can be achieved. What is at first surprising, and must have excited artists developing the ideas, is that all the other diagonal families are forced to meet at their own vanishing points, and that these all lie on the horizon.

(i) A base line with five points equally spaced

(ii) Line projected from each of the points on the base line passing through a single point – the vanishing point. (This is because in the grid these lines are parallel and at right angles to the plane of the page.)

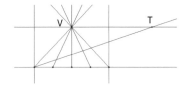

(iii) Draw the diagonal from the front left corner of the grid to pass through T, a point on the horizon: the horizontal line passing through the vanishing point V.

(iv) Where this diagonal line cuts the vertical line through the front right corner of the grid identifies the distance of the back edge of the grid (shown by the additional horizontal line here).

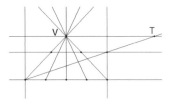

(v) Where the diagonal from front right to back left of the grid cuts the grid lines that converge at the vanishing point defines the positions of the horizontal transversals.

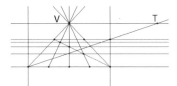

(vi) The grid can now be constructed.

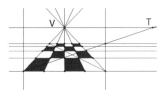

Figure 4.3f

In the next task, you are asked to follow instructions to construct a pavement from a given viewing position.

Task 4.3.1 Constructing a Projection Challenge

Using Figures 4.3f, g and h, follow the instructions to construct a projection of a pavement.

1   Draw the front of the grid AB, including the five equally-spaced points.
2   Mark the mid-point of the front edge M.
3   Mark the viewing position O.
4   Mark the vanishing point V, and join V to each of the five points on AB.
5   Mark point V' where the vanishing line drawn through V meets the vertical line through B.
6   Mark point O' where length V'O' = length OM.

7   Join A to O' and mark H where AO' intersects vertical through B, and M' where the horizontal through H intersects the line MV.

8   Look at Figure 4.3h. The right hand side shows a side view through MM'.

9   OM' gives a line of sight from observer to M'. This cuts the vertical through M at H' and gives the back edge CD of the grid.

10   This allows the diagonal BD to be drawn in the perspective drawing.

11   Where BD intersects the lines produced to the vanishing point, draw the grid's transversals.

12   Distance OM = O'V. Since triangle O'V'H is congruent to OVH', OV =O'V' = distance from plane of the drawing to observer = VV".

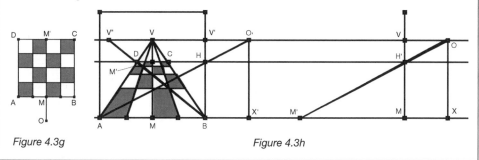

Figure 4.3g   Figure 4.3h

Comment

The method allows you to find where the artist expects you to stand when viewing a painting. Making sense of the step-by-step instructions is no easy matter, which might make you sympathetic to the struggles of artists in the sixteenth century. You may have drawn on geometrical knowledge that you had to revisit from earlier parts of this book to make sense of the statement 'OV = O'V' since triangle O'V'H is congruent to OVH' '.

The geometry developed in the preceding task allows you to determine the best viewpoint for a painting. In the next task you will have an opportunity to do this using a painting of the Interior of Antwerp Cathedral (1617) by Peter Neefs the Elder (Figure 4.3i). V"

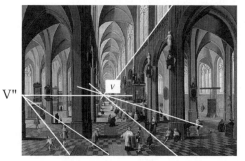

Figure 4.3i

First, the vanishing point, V, is found using the parallel lines of the building. The point V" is found by extending a diagonal of the grid of tiles on the floor and finding where it intersects the horizontal line through V. This allows the viewing distance V"V to be determined – this is the distance you should stand away from the canvas, with your eye level with V, to get the best possible view of the painting.

Now you can work out how far from the painting the artist wanted you to view it to get the most realistic impression.

Of course the reproduction here is much smaller than the real thing, which is 71.5 cm by 105 cm. From this information you can work out the viewing point for the real painting.

Reflection 4.3

Did you skim over the detail of this section without really engaging with the technicalities? What would motivate you to go back and master the calculation of ideal viewing point so that you could use it the next time you see a Renaissance painting?

4.4 LOCI: PATHS OF POINTS

Having traced rays of light from object to eye via a picture plane, you are now invited to shift your attention to tracing the paths of points moving under specified constraints, usually to do with a linkage or other mechanism. If you trace the motion of a point that is constrained in some way, you get a path called a *locus* (Latin for 'place').

You will find using a dynamic geometry software package invaluable in working through many of the tasks in this section – an alternative will be to sketch or draw accurately the situation, and to complement this activity by a thought experiment attempting to imagine what is happening.

Task 4.4.1 First Imaginings

Imagine …

… a circle …

… a point that is free to move on the circle …

… a fixed point outside the circle …

… a piece of elastic joining the two points …

… a knot at the mid-point of the elastic.

What is the path traced out by the knot as the free point moves around the circle?

What is the effect of changing the position of the fixed point? What happens if the circle is replaced by a square, a triangle, a quadrilateral?

Comment
There is an interactive file related to this task called '**4f Locus**'. It can be very helpful for checking the images that you had, but if you go straight to working with software before making conjectures you may lose some of the power afforded by the dynamic geometry software.

Some of the possible variations in the task are overtly available in the file: changing the size of the circle, the position of the fixed point and the position on the elastic of the knot. Once you see what is going on, you can imagine and express a generalisation that allows the circle to be any curve whatsoever!

Working on mental imagery is very useful for helping learners develop their power to imagine and to express what they imagine in movement, pictures and words. Inviting learners to say what they see in their minds, so that others can try to see what they are seeing leads to vibrant discussion and negotiation of interpretations. It also provides a model for work on other objects that are not mental: diagrams, figures, even sets of exercises or complicated formulae.

A rich development of Task 4.4.1 is to allow both ends of the elastic to be points moving round circles. There are a great many aspects that can change, such as: which direction the two points are going round their respective circles; the different speeds at which they are each travelling; their relative starting points; and, later, the position of the knot other than at the mid-point. Consequently, it is necessary to make choices and to try to be systematic. It may also be helpful to turn the task around and ask in how many different ways you could get a circle as the trace, or whether it is possible to get familiar shapes such as a straight line segment.

Locus problems arise whenever there is a moving mechanism, such as complicated hinges, sliding ladders, crank-sliders and ironing boards, to name just a few.

Task 4.4.2 Sliding Ladder

Imagine a ladder propped against a vertical wall with its foot on smooth horizontal ground. What path does its mid-point trace out as the ladder slips to the ground?

Imagine before you sketch; sketch a conjecture before you use dynamic geometry software.

Comment
To build a suitable figure you will need to specify a segment for the ladder length, then work out a way of displaying the ladder when the foot (or the top) of the ladder is in a specified position. Then you can move that position and see what happens!

Engineers designing the first engines needed to be able to convert rotational motion into linear motion and vice versa. A mechanism to do this is called a *crank slider* and is used on steam engines today.

The interactive file '**4g Crank Slider**' shows a point P that can move around a circle with centre O. Attached to point P is a rod of fixed length that can slide so that its other end P' travels along a straight line – in this case the line is horizontal and passes through O (see Figure 4.4a.).

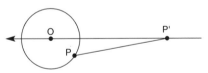

Figure 4.4a

Task 4.4.3 Crank Slider

Work out how the motions of P and P' are linked. What are some of the possible variations in the setup? Imagine and predict the shape of the graph of the distance of P' from O.

What is the path of P' if it is constrained to move vertically rather than horizontally?

Then experiment with the interactive file '**4g Crank Slider**'.

Comment
Among the things that could be varied are the radius of the circle, the direction of motion of P, the speed of rotation and the distance PP'. There are constraints however, because under some conditions the crank will get stuck, or cannot move at all!

When there are many things that could be changed it is helpful to ask yourself what features of the output change as a result of systematically varying various features of the slider.

Motion of the type you have been investigating here is important as it involves oscillations. You may have met such motion elsewhere in mathematics or science – for example, in the branch of mathematics often called 'mechanics'. Notice that in this task you considered varying certain aspects of the situation; it is still possible that certain aspects of the motion remain invariant. For example, the motion of P' is always oscillation: its amplitude is the maximum distance it travels from a central point, it is moving most quickly as it passes through the central point, and it comes to a halt briefly at each end of the motion.

Another common example of oscillation is the oscillating spring.

Task 4.4.4 Spring Time

Consider a length of elastic that has a mass hanging on its end so that it has an initial stretched length AB.

Now imagine that the mass is further stretched by an amount c and released.

What can be varied in the set up of this situation?

The resulting motion can be explored using an interactive file such as '**4h Oscillation**'. Notice that the oscillating motion is achieved using the projection of a point travelling around a circle.

Comment
Some of the aspects of the task that you can vary are the initial length of the elastic and the amount by which you stretch it. What stays the same and what varies as you alter features of the simulation?

To end this section, think for a moment about a conventional ironing board (Figure 4.4b). Like many familiar objects, there are interesting questions lurking!

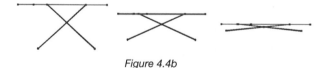

Figure 4.4b

Task 4.4.5 Ironing Board Design

What is the locus of a foot of a leg as the ironing board collapses onto the floor? What about the midpoint of one of the legs?

How did the designers choose the hinge point where the legs cross, and the point on the board itself at which one leg is hinged? What happens if you try changing these?

Comment
When you have considered this mentally, experiment with the interactive file '**4i Ironing Board**'.

Reflection 4.4

What advantages and disadvantages have you encountered in using dynamic geometry software to explore movement of physical objects?

Set yourself to look out for three different mechanisms, and see what sorts of locus questions you can ask about them. Mid-points are usually an interesting starting point!

4.5 PEDAGOGIC PERSPECTIVES

Many of the tasks and topics in this chapter have been inspired by or stimulated by design issues, whether in engineering or in painting. Using mathematics to (re)present spatial and numerical relationships is usually described as *mathematical modelling*. The idea is to build a mathematical system that models or captures the essence of the important relationships, making it possible to try things out without going to the expense and danger of actually making physical versions until the design appears to be safe and effective.

Geometry, in the form of distilling and expressing spatial relationships, and making use of things that must always happen (expressed in geometrical theorems) is sometimes the mathematical language for modelling, and sometimes contributes to the world of imagery in setting up a symbolic model to be solved algebraically. Even the very first task of the chapter can be seen as an invitation to participate in some aspects of modelling: imagining and expressing what solid shapes are swept out by a rotating two dimensional shape.

Pedagogic Strategies

Say What You See

Asking learners to 'say what you see' is a useful strategy for getting them to articulate what strikes them in one or more objects, whether diagrams, pictures, physical objects, sets of exercises, or formulae. 'Say what you see' means exactly that, without elaboration or theorising, but pointing (verbally or physically) to features so that others discern things they might not otherwise have noticed. Several of the 'Imagine' tasks in this chapter are based on 'Say what you see'. This strategy highlights learners' abilities to articulate what they are thinking, and to negotiate terms and descriptions in order to reach agreement. One version of the strategy is to give each learner a card with a diagram on it, to be described to the others so that they can draw it for themselves without actually seeing what was on the card.

Reversing or 'Doing and Undoing'

Whenever you find yourself 'getting an answer', it is helpful to check your appreciation and understanding by reversing the problem, interchanging what you are given and what you are to find or show. Other variants of this strategy include characterising all those problem situations that give rise to exactly the same answer, and characterising all possible answers to similar tasks.

Pedagogic Constructs

Task 4.2.1 offered you the opportunity to experience Bruner's modes of representation: enactive, iconic and symbolic. That is doing things, in this case possibly handling objects (enactive), working with three-dimensional mental images of these objects (iconic) and translating these into 'conventional' representations (symbolic). In this instance you may feel that the idea of symbolic is perhaps being stretched as you are not working with symbols in the algebraic sense; however, as you are constrained so that your representations must conform to particular conventions you are developing an understanding of a visual symbolic system.

Dimensions of Possible Variation and Task Design

Notice how in Task 4.1.1a aspects that could be varied were only briefly indicated, while the task structure (imagine a shape revolved around an axis) remained invariant. Getting learners to suggest what could be varied involves them in making choices and hence in engaging more fully with tasks. A 'solution' to a problem is rarely of value unless you appreciate how it depends on features that can be varied or what we might call 'dimensions of possible variation' in the original problem. Things cannot usually be changed arbitrarily, so it is also important to consider the range of permissible change of each of the features that can be changed. The words *possible* and *permissible* are included as a reminder that learners and teachers may have quite different dimensions and ranges in mind.

For example, the commentary on Tasks 4.2.2 and 4.4.1 suggested several dimensions of possible variation, and implied that most, if not all, tasks are not only individual tasks but also gateways to worlds of exploration, with many possible modifications in order to suit the interests, propensities and needs of different learners.

Learners could be asked to construct for themselves two different solids with one or two elevations in common, with or without Unifix© cubes or some other physical objects, but they could also be shown a pair of objects with two elevations in common, and then invited to construct their own. Learners could be constrained to a specified number of cubes and asked to construct as many different objects as possible ('different' meaning perhaps differing in at least one, or at least two, or even at least three different elevations).

Choices have to be made between the many possible variations in task specification. The choice is governed by the aims and purpose of the tasks. The explicit (outer) tasks indicate what learners are to do, though there may be deliberate ambiguity in order to give learners opportunity to make choices for themselves. Inner aspects of tasks include mathematical themes to be encountered, such as reversing ('doing and undoing'), invariance in the midst of change, or freedom and constraint. They also include learners experiencing the use of their own powers such as to imagine and to express that in various forms, to specialise and to generalise, and to conjecture and convince. Inner aspects include pedagogic experiences such as developing a sense of some concept, technique or way of thinking through manipulating objects and repeatedly trying to articulate the sense until it becomes succinct and familiar, ready to be used in future manipulations.

In order to inform choices about how tasks are presented, it is important to become aware of your own propensities, so that you do not unwittingly force learners to fit in with your own style of thinking. Varying your approach is more likely to draw in all learners as well as providing opportunities to discover hidden strengths and to strengthen underdeveloped powers. For example, sometimes you can use objects to introduce ideas, and sometimes you call upon mental imagery; sometimes you start with particular cases, and sometimes you start with something more general; sometimes you start verbally, sometimes visually and sometimes kinaesthetically; sometimes the object is to design or 'make' a product, and sometimes the aim is to reach a level of understanding which enables learners to convince each other that they understand.

Some people have a definite preference for practical, familiar material world situations for motivating learners, while others motivate by getting learners to use their powers to imagine and abstract. These can be combined by varying when and how material and abstract versions are used. Hans Freudenthal in the Netherlands introduced the notion of Realistic Mathematics Education, which goes beyond what learners have already experienced, to make use of situations which can 'become realistic' for learners. If school does not challenge learners to go beyond their immediate experience, then it is failing in its aims, but if it fails to make contact with learners, then it cannot hope to involve and engage them fully and effectively.

Reflection 4.5

If 'getting learners to suggest what could be varied involves them in making choices and hence in engaging more fully with tasks', what stops teachers from doing this?

Introduction to Block 2

In this block, geometric ideas are met in the context of measurement.

Chapter 5 returns to the root meaning of *geo-metry* as 'measurement of the earth'. From farmers staking out land before and after floods by the Nile, to astronomers trying to chart the heavens, geometry was driven by practical questions. The chapter includes work on co-ordinates and trigonometry.

Chapter 6 focuses on language and points of view in the context of measurement involving two and three dimensions. Circles are used to form various solids and, in order to measure areas and volumes, circles are approximated by polygons. Nets are used to help visualise shortest paths on solids.

Chapter 7 is about the nature and role of reasoning in geometry. Pythagoras' theorem, in-circles and ex-circles are the basis of exploring a range of approaches to proof.

Chapter 8 is about how geometric thinking can assist in the visualisation of mathematical ideas and, in particular, the development of the concept of locus to gain a deeper understanding of graphs.

5 Invariance in Measurement

This chapter considers geometry from the point of view of measure, turning to the root meaning of *geo-metry* as 'measurement of the earth'. Although developed in ancient Greece, classical geometry was not in their school curriculum. It was used both by philosophers to inform their reasoning, and by astronomers. From farmers staking out land by the Nile before and after floods, to astronomers trying to chart the heavens, geometry was driven by practical questions.

In order to think about measure, it is necessary to recognise that there are two types of measure: practical and pure. The first section introduces practical measure for length, angle, perimeter surface area and volume. The second section introduces pure measure as co-ordinates, because this is where modern mathematics parted company with traditional geometry. The third and fourth sections develop trigonometry (literally *tri-gon measure*) as a measure of triangles, particularly of angles.

5.1 MEASURE AS INVARIANT

Practical measure uses conventional units such as metres and its derivatives, kilometres and millimetres, light years and nanometres. But since a measurement is done by comparison with a standard measure (once a bar of steel, now a vibration of a caesium atom), there is always an error involved. Strictly speaking, a measurement has three parts: a number (of units), the units involved, and an error interval within which the quantity is confidently placed. Pure measure compares two lengths exactly in an ideal figure. No error is involved. Ratios arise in geometry and arithmetic precisely because all measurement involves ratio comparison.

Task 5.1.1 Measurement Associations

Imagine considering moving a bookcase from one room to another but not being sure if it would fit through the doorway. If you did not have a ruler, how might you find out without actually moving it? What then is the essence of measurement?

Comment
Did you think of using some string, or even a broom handle?

The key aspect of any measuring is to make comparisons. One way to do this is to decide on a unit and to have some way of replicating the unit (or breaking it up into smaller sub-units) and juxtaposing this with the thing to be measured.

This is not always as easy at it seems at first. When measuring length with a rigid ruler, care is needed to start from the zero and to read the scale correctly. When string or a flexible measuring tape is used, additional care is needed to make sure that the device is taut.

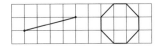

When a segment is displayed on a grid, learners sometimes attend only to the displacement in one direction, and so incorrectly judge the length of the segment in Figure 5.1a as four squares, and the

Figure 5.1a

perimeter of the octagon as eight. When measuring curved lengths, things become more difficult because the measuring device has to match the object as accurately as possible to get a reasonable measurement. Measuring circumferences of circles with string involves measuring the string held with the same tension against a ruler and necessarily involves more error than measuring straight lines.

Length, area and angle, like any measurement, assign a number that is a proportion or fraction of a specified unit. The number assigned is actually a ratio. For example, a stated length of 3.5 cm means that the ratio of the length of the object to the standard 1 cm length is 3.5, within a stated error band. For an area of 6 cm^2, the ratio of the area to the standard unit area is 6. For angle, the ratio of the amount of turn to a full turn is divided into 360, or 2π or 400 sub-units (to give degrees, radians or grads, units of measurement of angle used by different communities).

Assigning a number is itself problematic because measurement involves error. Measurements are necessarily estimates rather than exact. For example, a height of 174 cm suggests a height between 173.5 cm and 174.5 cm. but a person's height can vary a little between starting and finishing a train or car journey, and between morning and afternoon. As numbers, 4.2, 4.20, and 4.200 are different names for the same thing, but as measurements they are very different, because they have different implied precision: 4.2 m may be taken to mean somewhere between 4.15 m and 4.25 m, or 4.2 $\pm$ 0.05m. In order to present the same precision, but write the measurement in mm it would be necessary to write 4200 $\pm$ 50 mm, or 4.2 $\times$ 10^3 mm. In many contexts, especially engineering, knowledge of the precision of the measurement is vital.

When calculating with measurements it is common to produce many more digits than the original precision can justify. For example, if the lengths of the sides of two cubes were 1.2 m and 1.200 m, their volumes would be calculated as 1.7 m^3 and 1.728 m^3 respectively.

Relative Measurement

When developing a sense of measuring, pupils start by comparing and use phrases such as 'I'm older than her', 'I'm taller than him', 'Jim is heavier than his sister' and 'that looks bigger than this'.

Deciding which of two segments is longer tends to be easier to do visually when they are parallel, but much more difficult when they are at an angle.

Task 5.1.2 Estimating

Hold a ruler up horizontally and try to estimate how many copies are needed to make the height of a wall; hold a ruler vertically and estimate how many copies are needed to make the length of a room. Check your estimates by measuring.

Without using the ruler, estimate the thickness of this book. State a thickness it must be less than and a thickness it must be greater than. Now look at the ruler again. Without putting the ruler against the book, refine your upper and lower bounds so that they are closer to the actual measurement.

Without measuring, put the segments in part (i) of Figure 5.1b in increasing order of length. Why might some learners think that all three segments in part (ii) are the same 'length'? For each of the three pairs in part (iii), decide whether the horizontal segment is longer or shorter than the other.

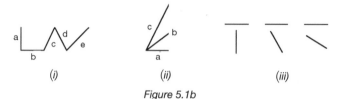

(i)   *(ii)*   *(iii)*

Figure 5.1b

Comment
Suddenly estimating becomes quite difficult, because you have to imagine rotating the unit in your head to match the object being measured.

A common misconception occurs when learners assume that if the beginnings and endings of a pair of segments are aligned vertically, then the segments are the same length: they are attending to the horizontal displacement rather than to the segment itself. Obliqueness some-times suggests greater length even though all the segments on the right are the same length.

Angles are notoriously hard to compare because of the misleading contribution made by the lengths of the arms. All the angles in Figure 5.1c represent the same rotation.

Figure 5.1c

Areas are even harder to estimate, and this is the subject of the next task.

Task 5.1.3a Estimating Areas

Which of the three regions in Figure 5.1d has the largest area?

Comment
In fact they are all the same, because the circles were made so as to have areas in the ratio of 1:2:3.

Figure 5.1d

It is bad enough estimating when the shapes are the same, but when the shapes are different, it is even harder!

Task 5.1.3b Estimating Areas Again

Draw a triangle. Draw a square outwards on each side of this triangle. Join the outer corner of each square to the nearest outer corner of the next square to form three additional 'flanking' triangles, each one between two adjacent squares. Which is the largest of the four triangles? Does it depend on your starting triangle?

> *Comment*
> Dynamic geometry software is excellent for this purpose, because you can vary the vertices of the initial triangle and watch what happens to the four areas. You may wish to experiment with the interactive file '**5a Four Areas**'.

Dynamic geometry software may convince you that there are some, non-obvious, invariants in this diagram, which are independent of the shape of the initial triangle But this is not a proof! The diagram from Task 5.1.3b and its areas will reappear later.

Area and Perimeter of Circles

Archimedes (287–212BCE) worked with areas, and with shapes such as those in Figure 5.1e. He determined that the area of the circle in part (i) of Figure 5.1d lies between the areas of the outer and inner square, and that the length of the perimeter also lies between the perimeters of the two squares.

(i)   (ii)

Figure 5.1e

Similarly, the area and perimeter of the circle lie between the corresponding measures of the inner and outer hexagons in part (ii) of Figure 5.1e. By calculating the areas of polygons with very large numbers of sides, up to 96, Archimedes was able to get a good estimate of the area and perimeter of a circle. He proposed that the ratio of the perimeter of a circle to the diameter was between $3\frac{10}{71}$ and $3\frac{1}{7}$. This ratio is denoted by the Greek letter π (pi).

He discovered that the area of a circle was a fixed ratio times the radius squared and that this ratio was also π. The perimeter of a circle is known as the *circumference*. Thus, for a circle, $C = \pi \times D$ and $A = \pi \times r^2$.

Relating Perimeter and Area

If you scale a figure by a factor of two in all directions (so all lengths double), the perimeter, being a sum of lengths, also doubles. The area will change by a factor of 4, the square of 2. Another way to see these relationships is to observe that the ratio of the perimeter to the length of some side or chord of a figure is invariant under scaling. So, for each figure, there is a perimeter-length ratio. Similarly, the ratio of the area to the square of the length of some side or chord is also invariant under scaling. In particular, every figure has an area-to-squared-perimeter ratio.

Task 5.1.4 Perimeter and Area Ratios

Find the ratio of the perimeter of a square to the length of the diagonal. Do the same for a regular hexagon and a regular octagon. (Use the longest diagonal.) Include the ratio for a circle – a *diagonal* is then a *diameter* and the perimeter is the *circumference*.

Find the ratio of the area of a square to the square of half the diagonal. Do the same for the regular hexagon and the regular octagon. Again, include the corresponding ratio for the circle if you know it.

Finally, find the ratio of the area of the shape to the square of their respective perimeters. You will find it helpful to record your results in a table such as the one shown overleaf.

SHAPE	PERIMETER (P)	LENGTH OF DIAGONAL (D)	RATIO P/D	AREA (A)	$(D/2)^2$	RATIO $A/(D/2)^2$	P^2	RATIO A/P^2
Square								
Regular hexagon								
Regular octagon								
Circle								
Other								

Comment

Scaling a figure leaves invariant the ratio of the area to the square of the perimeter or, indeed, to the square of any other measure of length such as a radius, side length or diagonal. If you are not sure about this, try calculating the ratios for different sized shapes.

The ratio of the area to the squared perimeter is 16 for the square, about 13.86 for the regular hexagon, about 13.32 for the regular octagon, and for a circle it is 4π which is about 12.57. Notice how the ratios are getting smaller as the number of sides increases. The circle has the smallest such ratio of any shape.

Perhaps the most amazing feature of a circle is that the ratio of perimeter to diameter, and the ratio of the area to the radius squared are the same, whereas for other shapes the corresponding ratios differ. You will work on this again in Chapter 6.

Changing the shape does not always change the area, or the perimeter:

Task 5.1.5 Fixing and Changing

Is it always, sometimes, or never, true that:

For any rectangle of a given area, there is another rectangle with the same area but larger perimeter.

For any triangle of a given area, there is another with the same area and larger perimeter.

For any rectangle (triangle) with given perimeter, there is another rectangle (triangle) with smaller area.

Comment

Did you draw a diagram or use apparatus such as string and squared paper? How did you arrange to hold the area constant yet alter the perimeter, or vice versa? Were you satisfied with a few cases or did you look for some general justification?

Did you think to consider what range of change is permissible for the shapes involved?

How did you decide whether the statements were always, sometimes, or never true? Did you try some examples first, or did you go to a direct proof? Most learners try examples first.

Breaking Up Shapes

Lunes are shapes formed by removing part of a circular disc from another circular disk, so-called after the various shapes of the visible moon as the almost-circular shadow of the earth moves across it. There are four shaded lunes in Figure 5.1f. The next task involves the area of a 'lune'.

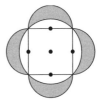

Figure 5.1f

Task 5.1.6 Lunes

Describe Figure 5.1f in words. Find the total area of the four shaded 'lunes': do not substitute an approximation for π, just use the symbol.

Comment
One example of a description of the figure is that each 'lune' is bounded by a half-circle on a side of the square and an arc of the circumcircle of the square.

 Were you uncertain where to start when you did not know the length of the side of the square? Did you assume it could be called *x* or l? When your answer did not contain π were you surprised? Did you check your working to find a slip, because in fact the total area of the lunes is the same as the area of the square and it does not involve π.

Every theorem contains a surprise, but surprise depends on having expectations. Here, the implicit assumption that an area involving a circle must involve π can be the source of a surprise. Further examples can be found in Wills (1985).

 To end the section, here is a small quotation about the power of geometrical thinking from 'The Clouds' by Aristophanes (324 BCE)

 STREPSIADES: What about these [surveying instruments]?

 STUDENT: Geometry.

 STREPSIADES: What can you do with them?

 STUDENT: Measure land.

 STREPSIADES: You mean my allotment?

 STUDENT: No, the whole world.

Reflection 5.1

In what ways does measurement depend upon invariance?

5.2 CO-ORDINATING WITH NUMBERS

Suppose you want to arrange to meet someone at a particular place. You can use all sorts of information, but suppose you want to be economical in your instructions. You are likely to use something akin to a map grid reference. To specify a point on the surface of the earth, you can use latitude and longitude. To specify a place on a map, you can use co-ordinates.

Co-ordinates on a plane

Co-ordinates are associated with algebra and are first met in arithmetic as positions on a number line, which is a one-dimensional co-ordinate system. Pairs of co-ordinates are used on maps with references such as F7, and as (x, y) in algebra and geometry. Other co-ordinates are possible, each uniquely defining position, but in different ways.

Task 5.2.1 Navigating with Different Co-ordinates

Imagine being in a boat within sight of land and that you have to hand a range-finder (which measures distance to a landmark), a compass and a global-positioning receiver (which tells you latitude and longitude). In how many different ways could you locate your position on a chart?

Comment
Possibilities include: distances from two identifiable landmarks; latitude and longitude; distance from nearest point on the shore and from some other point; angle between two landmarks and distance from one of them; angle between one landmark and two others. In each case if you took an extra measurement, it would be useful as a check on accuracy of measurement.

These methods illustrate different co-ordinate systems on a surface:

- With a range-finder and compass, you can use a polar co-ordinate system, which has an origin (pole), and measures position by distance to the pole and angle from a fixed direction through the pole (for example, north).
- A bipolar co-ordinate system uses distances from two fixed points (poles).
- A Cartesian co-ordinate system uses two fixed axes, and measures distances parallel to those axes (usually perpendicular). On the chart, longitude and latitude may appear to be perpendicular co-ordinates even though in fact they are curves.
- Two bearings (compass readings) for landmarks is a bi-angular co-ordinate system.

Thinking about the situation of the boat, and about different co-ordinate systems is easier if you have experience of maps and map reading; if learners do not have experience on which to draw, it is vital that they enact it physically on a model. Whenever sense seems to be slipping away, it is useful to try to regain confidence by finding something to do which makes the situation more concrete, more manageable and friendly.

The point of simplification and concrete examples is to gain confidence and to work out what is going on more generally. Talking about what you are doing helps you to clarify; recording what you are saying, whether in words, pictures or symbols is helpful as a way to become familiar with formal symbols and notation.

Task 5.2.2 Square Vertices

Starting with a line segment whose vertices are (a, b) and (c, d), find the co-ordinates of the other vertices that form a square with this segment as one edge.

Comment
Having drawn the segment and one of the two squares using it as one edge, you may find yourself a bit stuck. Since squares with edges running parallel to the axes are much easier to

deal with, try drawing a surrounding square that has each of the vertices of your square on its sides, but has its sides parallel to the axes as illustrated in Figure 5.2a.

Now observe some congruent triangles, and use the sum and difference of co-ordinates to work out the remaining co-ordinates.

Figure 5.2a

If the co-ordinates of the vertices of original line segments are all integers, then the co-ordinates of the vertices of both squares are likewise all integers, which might be at least a small surprise! This fact can be used on the diagram you drew in Task 5.1.3b.

Task 5.2.3 Integer Vertices

Draw a triangle whose vertices have *integer* co-ordinates (dotty paper is ideal for this purpose). Now imagine a square drawn outwards on each side (as in Task 5.1.3b). Will all the vertices of the squares have integer co-ordinates? (Conjecture, draw some, check your conjecture, then think how you would convince someone else.)

Comment

To collect some evidence you probably worked by trial and error on a few cases, but these can only be suggestive. They provide evidence but not proof. A justification of the conjecture must cover all possibilities, and a way of doing that is outlined below.

The vertices of a triangle on a co-ordinate plane are not usually integers, so if you are exploring something about *all* triangles and the co-ordinates of their vertices, it is necessary to use co-ordinates such as (x, y) or (x_a, y_a) or (x_1, y_1), where the letters denote any permissible number, whether restricted to integers or allowed to be any real number.

To denote three general vertices there are some simplifications possible. The origin of the co-ordinate system can be anywhere you choose, so it could be taken to be at one vertex. The first axis can be in any direction, so it can pass through a second vertex of the triangle, and this axis can be scaled arbitrarily, so the first vertex can be used to determine the unit of measurement. You now have two vertices, $(0, 0)$ and $(1, 0)$ for the triangle. The third vertex, which has to be general, can now be denoted by (x, y). Interestingly, full generality for a triangle is achieved with just two variables.

Using dynamic geometry software, the third point can be dragged about, but dragging does not produce a proof that some relationship will always hold, rather it is a tool for trying a large number of cases quickly. In fact, you can only ever consider a finite number of cases even with dynamic geometry software, and the accuracy of calculation within software may occasionally suggest an incorrect conjecture. Geometrical thinking involves reasoning about *all possible* cases, not just a few particular instances.

Task 5.2.4 Revisiting Task 5.1.3b

Draw a triangle and label the vertices with the co-ordinates (0, 0), (1, 0) and (x, y). Draw a square outwards on each side of the triangle. Join each square to its neighbour by joining nearest vertices.

Find the co-ordinates of the other vertices of each of the squares. Find the areas of all four triangles.

Comment

Three of the triangles have a side on the square of side length =1; use these sides as the base of each. Also, all have the same height, so all three have the same area. To deal with the fourth triangle, imagine another pair of axes with their origin at the point (x,y) and one axis along the side of the original triangle to (1,0). Then the fourth triangle and two others all have the same length base (the sides of another square), and the same height and so must have equal areas. Therefore, the areas of all four triangles must be equal.

This method of finding areas of triangles with given co-ordinates can be generalised. Starting with a triangle with vertices (x_1, y_1), (x_2, y_2), (x_3, y_3), the area can be found by embedding the triangle in a rectangle, as shown in Figure 5.2b, and then using appropriate right-angled triangles to subtract areas.

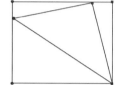

The resulting general formula

$$A = (x_1y_2 + x_2y_3 + x_3y_1 - x_1y_3 - x_2y_1 - x_3y_2)/2$$

is useful if you are doing computations, for example in software.

Having worked on co-ordinates in a plane the next sub-section considers co-ordinates on the surface of a sphere; the Earth can be approximated by a sphere.

Figure 5.2b

Co-ordinates on the Surface of a Sphere

Imagine drawing a circle around the surface of a perfectly spherical Earth so that it has the maximum circumference possible. There is of course an infinite number of these; perhaps you chose a special case such as a circle passing through the North and South Poles, or one that passes around the Equator. Such circles are called great-circles: aircraft flying between two cities follow *great circles*. This is because this is the shortest route.

To identify the position of a point on the Earth we use two sets of circles, as shown in Figure 5.2c.

Figure 5.2c

Meridians of longitude – half great-circles passing through the Poles. One of these – the Prime Meridian – is designated as being at 0° (passing through Greenwich, England) and any other is then measured as being so many degrees east or west of this.

Parallels of latitude – these are circles parallel to the Equator and are consequently decreasing in radius and circumference towards the Poles. Each can be identified by measuring the angle between the radius from the centre of the Earth to a point on the parallel and the radius from the centre of the Earth to a point on the Equator on the same meridian. Using such a system it is necessary to indicate whether the angle is North or South of the Equator.

Since it is possible to give unique measures of longitude and latitude for each position on the surface of the Earth, latitude and longitude provide a *co-ordinate* system for the surface of the earth. A point on the surface of a sphere is uniquely determined by two angles, one measured by a meridian great circle and the other from the equator. To co-ordinate space with this system, you include a third number that determines the radius of the spherical shell centred at the origin on which the point lies (see Figure 5.2d).

Figure 5.2d

Reflection 5.2

What systems of co-ordinates are you used to using to locate a position precisely? What are the advantages and disadvantages of different systems?

5.3 MEASURING ANGLES

The word trigonometry has many similarities with *geo-metry*, as it is made up of *trigon* (triangle), and *metry* (measurement). Astronomers needed to able to calculate all the sides and angles of a triangle given only partial information, and this gave rise to trigonometry. It all starts with the measuring of angles and the use of Thales' theorem that you met in Task 2.4.1. Before you continue, open and animate the interactive files '**5b Trig1**' and '**5c Trig2**'.

Suppose you want to achieve a measure of the angle *A* in Figure 5.3a.

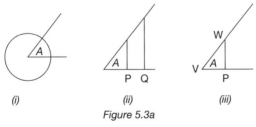

(i)   *(ii)*   *(iii)*

Figure 5.3a

A direct method is to insert a circle radius 1, centred at the vertex, and then use the length of the arc of the circle as the angle measure. This is known as radian measure of an angle. Since an arc length of 1 gives an angle of about 57.3°, the radian values for commonly used angles are not whole numbers. However, as full turn for a circle of radius 1 would be an arc of 2π, this gives 2π as the measure of a full turn, the equivalent of 360° in degrees.

Circular arcs are hard to measure when you are out in the field. Someone noticed that if you draw a right-angled triangle as shown in Figure 5.3a (ii), then using Thales' theorem, the ratio of any two of the sides of the triangle depends *only* on the size of the angle, and not whether the point is P or Q or any other point (not the vertex of course!).

Trigonometric Ratios

However, there are a great many ratios, each of which is invariant when the position of P is chosen. From Figure 5.3a (iii), there are six ratios:

$$\frac{WP}{VP}, \frac{WP}{VW}, \frac{VP}{VW} \quad \text{and} \quad \frac{VP}{WP}, \frac{VW}{WP}, \frac{VW}{VP}.$$

Of course these are interrelated, (for example, the last three are the reciprocals of the first three) and this is the source of both the power and the complexity of trigonometry. Just as it is convenient to give names to numbers that are hard to specify precisely in any other way (such as $\sqrt{2}, \sqrt[3]{4}, \pi$ and so on), it proves to be most convenient to give names to these ratios as measure of the angle. The names arise from Latin: *sine* from *sinus* = *curve*, *cosine* from *co-* for complement, *tangent* = touching, and *secant* = cutting.

Figure 5.3b

Since the size of the right-angled triangle is permitted to change, and it is the ratio of lengths of sides that is of interest, it makes the calculations easier if you take the hypotenuse to be of length 1 unit, as shown in Figure 5.3b. (Notice that the angle here is denoted by the Greek letter θ (pronounced theta). You will find this convention used quite often in geometry texts. The mixture of items of Latin and Greek origin in mathematics texts written in English goes back to the first English text which was written by Robert Recorde, physician to Queen Mary Tudor.)

You can think of the triangle sitting on graph paper with the bottom left placed on the origin. The co-ordinates of the other two vertices of the triangle can then be determined by the cosine and sine of θ.

Figure 5.3c is another representation of the trigonometric ratios. This time, the sides of the right-angled triangle have been changed by dividing the length of the sides of the previous triangle by cos θ, so that the side adjacent to the angle is of length 1. This diagram is helpful for illustrating the tangent (tan θ) and secant θ (or sec θ).

Figure 5.3c

Task 5.3.1 Seeing Ratios

Carefully inspect Figure 5.3d.

Assume that the length OP is one unit. Using your own labels for the other points, pick out all the triangles that include an angle θ.

Then describe using your own words and labels the meaning of each of the terms represented in Figure 5.3d.

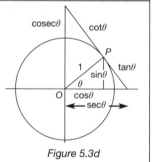

Figure 5.3d

Comment
You will have noticed, if you did not know it already, that $\sin \theta = 1/\csc \theta$, $\cos \theta = 1/\sec \theta$ and $\cot \theta = 1/\tan \theta$.

'Solving a triangle' means finding any unknown sides and angles from known data about the length of sides and the size of angles.

If the triangle is right-angled, then you need to know the sizes of one other angle and one side, or of two sides to determine the rest of the sides and angles. The procedure is to draw the triangle that you are trying to solve, and compare it with one of the triangles in Figure 5.3b. The key to solving right-angled triangles is being able to find the scale factor that maps one of these triangles onto the triangle with unknowns. Having found the scale factor, you then compare the sides and use a calculator to find the unknowns.

A related historic problem that the ancient Babylonians solved was finding the radius of the Earth. Their approach is well documented (see, for example, St Andrews webref). It is also possible to find an approximate value of the radius in other ways.

Task 5.3.2 Measuring the Earth

Standing on the beach, a yacht with a mast 12 m high is observed sailing over the horizon. The observer, whose eyes are 2 m above sea level, is talking to one of the crew on a mobile. As the boat disappears, the observer asks, 'How far are you from the beach?' The sailor replies, 'We have just passed Flounder Point on the Island, so, according to the chart we must be about 11 km from the beach.' Using these results, what would be an approximate value for the radius of the Earth?

Comment
Make sure you draw a diagram. The radius of the Earth is approximately 6380 km.

How then might you measure angle in three dimensions? What could be meant? Imagine a sphere, and a region of the surface. Join every point of the region to the centre of the sphere. The result is a cone or cone-like figure. The 'space swept out' at the apex of the cone forms the *solid angle* specified by the region. The size of the solid angle mimics the definition of radian measure. Take any region on the surface of a unit sphere and find its area. The ratio of the area of the region to the surface area of the whole sphere gives a measure of the space swept out by the region, that is, of the solid angle.

Reflection 5.3

What changes in your sense of angle when you think of trigonometric ratios as measure of angles? Which angles have the same *sine*, the same *cosine*, the same *tangent*? Which have both the same *sine* and the same *cosine*?

5.4 CONNECTING TRIGONOMETRY AND CO-ORDINATES

So far, the trigonometric ratios have been considered in terms of a right-angled triangle without reference to any co-ordinate system, though there was a hint of a co-ordinate system in the interactive files '**5b Trig1**' and '**5c Trig2**'. As you will see in this section, it becomes important to be able to think about the sine and cosine of angles greater than 90°. In order to lay a firm foundation for extending the ratios in this way, it is helpful to think back to the unit circle.

In Figure 5.4a, the *x* co-ordinate of point P is cos *A*. The *y* co-ordinate is sin *A*.

Figure 5.4a

In section 5.3, the problem of *solving triangles* was described for right-angled triangles. This is a special case of the more general problem for general triangles. Any triangle, in its turn, can be broken into right-angled triangles, so in principle the problem is solved. 'To solve' means to find all the side lengths and angles (and where desired, diagonal lengths or other information about the figures). Any polygon can be cut into triangles. (Although this may seem intuitively obvious, it is not quite as easy to prove as you might imagine, because it is very difficult to state precisely, but generally, where to find the first triangle to cut off.)

Sine and Cosine Rules

However, people have had to solve triangles so often that more powerful formulae have been devised to make the process easier. For example,

the sine rule $a/\sin A = b/\sin B = c/\sin C$,
and the cosine rule $a^2 = b^2 + c^2 - 2bc \cos A$.

Both rules are derived by starting with a 'general' triangle and splitting it into two right-angled triangles. The next task starts by working on the sine rule.

Task 5.4.1 Deriving the Sine and Cosine Rules

Draw a non-right-angled triangle. Mark the vertices A, B and C. Mark the sides *a*, *b* and *c* opposite angles A, B and C. Turn your drawing if necessary so that the triangle is sitting on a base with an acute angle at the top. Draw in an altitude to cut the triangle into two right-angled triangles. Using each right-angled triangle in turn, write down an expression for the height of the altitude. Put these two expressions together and rearrange them to get one part of the sine rule. Use another altitude to get the other equality. Check an obtuse angled triangle case in which some of the altitudes meet the base outside the triangle.

The cosine rule with a^2, b^2, and c^2 is a general case of Pythagoras' theorem. Draw a triangle, label the sides and angles, draw in an altitude (say height *h*). Now write down an expression of Pythagoras' theorem for each right-angled triangle in turn – if the base is *c* you will need to label its two parts *x* and *c − x*.

For each of these two expressions, make h^2 the subject on the left-hand side. Now each of the right-hand sides of both expressions is equal to h^2 and hence equal to each other.

Writing this equality down gives you one form of the cosine rule. It can be written with cos A, cos B, cos C, a^2, b^2 or c^2, as the subject of the formula, depending on which two right-angled triangles you started with.

Comment
An alternative derivation of the cosine rule is given in Chapter 7. Rather than memorising derivations, it is enough to remember to discern some right-angled triangles, even if you have to introduce them yourself, and then make use of Pythagoras' theorem.

In deriving the sine rule you found an expression for the height of a triangle. This can give a formula for the area of a triangle: Area(Δ) $= \frac{1}{2} ab \sin C$, with two other variants found by changing which sides are used: Area(Δ) $= \frac{1}{2} bc \sin A = \frac{1}{2} ca \sin B$. These formulae are equivalent to $\frac{1}{2} \times$ height $\times$ base but the height is given in terms of other known facts.

Task 5.4.2 Triangles, Squares and Triangles Again!

Draw a triangle, draw a square outwards on each side, and join the vertices of the squares to get three more 'flanking' triangles. Show that each of the new triangles has the same area as the original triangle.

Comment
Label the triangle so that A, B and C are the sizes of the angles at A, B, and C. Label the opposite sides a, b, and c.

Label all the other sides that are equal in length to a, b or c. Mark in all the right angles and in each outer triangle, find the value of the vertex it shares with the original triangle. Now use the sine formula for area. The three apparently different expressions are actually all equal by using the sine rule. You may have to use the fact that sin $(180° - A)$ is the same as sin A.

Inverse Trigonometric Ratios

When using trigonometry to solve right-angled triangles, you often know two sides and need to find an angle. For example, if two perpendicular sides were both 3 cm long then you would find that tan $A = 1$. To find angle A, you would have to 'undo' this tan function, that is, use the inverse function. In other words, saying 'tan $A = 1$' is the same as saying 'A is the angle whose tan is 1'. The superscript in $\tan^{-1}$ is a way of saying 'an angle whose tan is'. You would often use a calculator or tables to find the angle using $\tan^{-1}$.

Task 5.4.3 Doing and Undoing

Assume the cells are squares in Figure 5.4b. Find tan A, tan B and tan C.

Use the inverse function $\tan^{-1}$ and a calculator or tables, find the sizes of angles A, B and C.

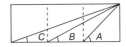

Figure 5.4b

It is interesting to note that $B + C = A$.

Noticing this relationship might lead you to a conjecture or even initiate an investigation – are there other examples where $B + C = A$? Check the diagrams in Figure 5.4c to see if $A = B + C$.

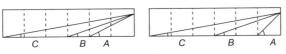

Figure 5.4c

Comment
In checking you found that in some diagrams $A = B + C$, but not always. In Chapter 8 you will see how the trigonometric ratios for sums of angles relate to the trigonometric ratios of the angles.

Care has to be taken when undoing trigonometric functions. Thinking of Figure 5.4a, there are many angles that have a tangent value of 1, for example 45°, 225°. It is possible to extend the notion of angle turned, and include even more possible angles such as 405° or ¯135°. The calculator can only give one value, known as the 'principal' value for the angle. For positive values of tan A, that is the angle between 0° and 90°.

Reflection 5.4

What more do you need to do to develop confidence and facility with trigonometry?

5.5 PEDAGOGIC PERSPECTIVES

Some of the ideas in the chapter may have been a review of familiar ground for you, but other parts may have been less familiar. Think about how a first exposure might, through additional experience, turn into mastery of a topic. For learning to be effective, it is vital that the learner takes opportunities to look at or work with a familiar physical object, to imagine and draw, and to symbolise or record in some manner. But this takes time. It is unreasonable to expect people to master something the first time they see it go by, and unreasonable to expect people to be able to write down what they are thinking before they have tried to say it to someone (even just themselves), and to modify that articulation until it becomes succinct and expressive.

It has already been recommended that you develop a notebook of ideas connected with different topics. Measurement is a topic that cannot simply be 'dealt with' in a sequence of lessons and then forgotten. It keeps coming up in different guises and in different contexts. What, then, are the key ideas?

Structure of a Topic

In preparing to teach there are many aspects of a topic to consider. We suggest that it helps to think about three strands: what learners will actually do, their motivation and the ideas they will encounter and work with.

The three strands can be exemplified by questions as follows.

Behaviour (what learners will actually do)

- What new vocabulary or language patterns do pupils need to learn?
- What related vocabulary are they already familiar with, perhaps in a different context?
- What procedures (or techniques and methods) do pupils need to master?

Motivation (emotion/affect)

- What problems does the topic help solve?
- In what other contexts does the topic or its ideas arise?

Awareness (the ideas they will encounter and work with)

- What images and ways of thinking are involved?
- How are learners' powers called upon?
- What mathematical themes and heuristics are involved?
- What misconceptions might some learners have and how might they be counteracted?

These ideas are discussed more fully in Chapter 13.

Without such a checklist, topics tend to be mostly about behaviour and teaching is impoverished. Many people find it helpful to use the three sections as a framework for collecting and developing notes on every topic. Every time you encounter something unexpected from a learner, or a different task that might promote some aspect, it can be added to the notes. The notes then make it easy to refresh your memory about the topic next time you teach it.

Task Design and Augmentation

Many tasks can be approached in different ways, and each usually has advantages and disadvantages. It is important for learners to appreciate that there is often more than one approach to a mathematical task, and when one method seems not to be working then it is sensible to try a different approach. Many traditional texts do not give this message.

You may have noticed that many of the tasks have invited you to imagine. The reason for this is that the power to imagine mentally is often underused and it is a very important power in thinking geometrically. Whenever you look at a diagram or figure, whether on a board, on a page or on a screen, it is important to be able to 'see beyond' the particulars to the generality that is implied. Every diagram depicts relationships. In order to pick up on those relationships you have to discern the key elements (segments, angles, triangles, circles), including shapes that may not yet be presented visibly in the diagram. A good way to develop this power is to invite learners to imagine, both before showing them something on a screen, and after the screen has been cleared.

One way to make tasks more effective is to find ways to phrase the task so that learners make choices (as in what counts as the same) and exercise their powers such as mentally imagining before actually doing. For example, a task that asks 'in how many ways … ?' tends to be more successful in getting learners to develop flexibility and to look for alternative strategies than a task that asks for a single answer.

Generalising is at the heart of mathematics. For example, every quadrilateral (and every polygon) can be split into triangles. The results you obtain may not be quite as elegant as those with the triangle, but the idea of seeking generalisations is important.

In section 5.4, you were concerned with making connections. This is important in mathematics and in all learning. One way to think about learning is to think about new ideas being developed in the learner's mind as they are stimulated by new experiences, and also how these ideas link with each other and with each learner's prior ideas. This is sometimes called 'building knowledge schemas'. From this perspective a schema might be thought of as a metaphor.

Piaget talked about assimilation, accommodation and rejection of concepts, and these can be thought of as three of the ways in which learners build schemas. Assimilation occurs when a new concept fits comfortably into the learner's existing schema, accommodation occurs when adjustments need to be made such as misconceptions being sorted, and rejection occurs if the concept seems not to fit and the learner rejects it. Not all concepts that do not fit the learner's schema are rejected; learners can have a number of disconnected schemas in their minds and therefore hold two contradictory concepts.

If mathematics is thought of as separate from other aspects of life, then mathematical schemas are likely to be separate from other schemas. If all aspects of life are looked at mathematically, then the schemas are more likely to be integrated.

One way of representing a schema is to use a concept map or mind map. These maps are usually drawn with 'bubbles' and with lines joining them. The bubbles can be thought of as labels for concepts while the lines can be thought of as the connections between them. More lines on the map might suggest more connections made and, perhaps, a richer understanding.

Now think about your pupils doing mathematics. Do they see you doing mathematics? Do they see you getting stuck, trying different methods, conjecturing and generalising, or do they only see you acting? Pupils learn a lot incidentally, and they learn from watching others; mathematics teachers therefore have a responsibility to demonstrate to pupils the delights and frustrations of the subject.

Although current ideas seem obvious and unforgettable, it is curious how quickly ideas and connections fade away. Making a written record actually helps to keep them available; teaching someone else is even more effective. The act of writing notes, when you treat the paper as 'teaching an absent friend' can be a valuable component of learning.

Reflection 5.5

Go back over the list of questions in the 'Structure of a Topic' section in the context of measurement, co-ordinates, and trigonometry for some class you will be teaching.

How are measurement, co-ordinates and trigonometry connected? Construct a mind map showing the connections you have seen.

6

Measuring Areas and Volumes

As with Chapter 2, this chapter focuses on language and points of view, but now in the context of measurement involving two and three dimensions. At the beginning of the chapter circles are used to form various solids, and in order to measure their areas and volumes, circles are approximated by polygons. Later in the chapter you will work on nets and pyramids

6.1 LENGTHS AND AREAS OF CURVED REGIONS

The area and perimeter of a circle were discussed in Chapter 5 and you worked on related ratios in Task 5.1.4, together with other shapes. You saw how Archimedes developed an approximate value for π by trapping the circle between polygons with an increasing number of sides, obtaining finer and finer approximations. A similar approach can be used to investigate the ratio of the area of the circle to the perimeter.

Task 6.1.1 Estimating Circles

Draw a circle. Now draw in two diameters at right angles to each other. Now bisect all of the angles at the centre with two additional diameters. Imagine repeating this process a number of times until you can only just see the individual sectors. (Figure 6.1a part (i) shows just one angle being bisected each time.)

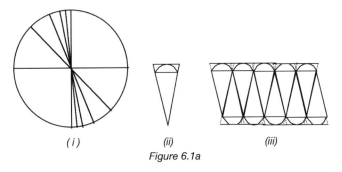

(i)   *(ii)*   *(iii)*

Figure 6.1a

Comment

The circular arc of each sector is very, very close to being a straight line, in fact, the arc is squeezed between the interior chord and the exterior tangent as shown in (ii) of Figure 6.1a – a single sector is shown with exaggerated arc so that it is visible.

Imagine the sectors rearranged so that alternate pieces are upside down as shown in part iii of Figure 6.1a. The figure then approximates to a parallelogram that gets closer and closer to being a rectangle as the sector angles are reduced.

The dimensions of the rectangle to which the parallelogram is converging are r (the radius of the circle), and $C/2$ (half the perimeter of the circle), respectively. The area of the rectangle, which closely approximates the area of the circle, is their product. Thus for a circle Area $= r \times C/2$.

This result can be used to demonstrate the truth of the statement from Chapter 5: 'Perhaps the most amazing feature of a circle is that the ratio of the perimeter to the diameter, and the ratio of the area to the radius squared are the same'.

Archimedes' method of trapping the curve between two families of polygons whose perimeters approach the same limiting value can also be used to find the lengths of other curves that can be approximated by polygons. Areas enclosed by curves are found similarly, by trapping the region between two polygonal regions, an inner one and an outer one.

Circular Paths

To see the principle in action, consider the following variation for circles.

Task 6.1.2 Circular Path

The circular path depicted in Figure 6.1b is to be 1 m across, made from paving tiles in the form of trapeziums. The radius of the middle of the path is to be 10 m. If there are to be 50 trapeziums making up the path, work out the dimensions of the slabs, the area of the path in slabs, and the area of the ideal annular path.

Comment

The area of each trapezium is average width × height. *Figure 6.1b*

Did you think to generalise the dimensions of the path? If you do, you will see that as the number of slabs gets very large, the total slab area will get closer and closer to the area of the annulus. If the circle through the centre of the path has radius R, and the path has width r, then the area of the annulus turns out to be $(2\pi R)r$, which is closely approximated by the total width of all the trapezia times their 'vertical height', namely their area.

The area of a sector of a circle can be found by thinking in ratios: the ratio of a sector whose angle is θ in radians, such as the one in Figure 6.1c is $\theta/2\pi$ of the whole circle, and so the area of the sector must be $(\theta/2\pi) \times \pi r^2$, which is $\theta r^2/2$.

Figure 6.1c

Lunes

To complete this section on curved areas, here are some somewhat surprising area calculations involving *lunes*, which you met in Chapter 5.

Task 6.1.3 Lunar Areas

The triangles in Figure 6.1d are both right-angled triangles, one with a vertex at the centre of a circle and the other with the diameter of a circle as hypotenuse. The lunes are cut from semicircles with their centres on the corresponding sides of the triangles.

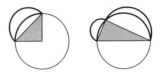

Figure 6.1d

Estimate by eye the relationship between the areas of the shaded regions and the lunes in each of the diagrams. After you have estimated, calculate the areas.

Comment

Areas of lunes can often be found by subtracting known areas from areas of semicircles or circles. In the first example, you can calculate the shaded area as half of r^2 , the white segment as ($\frac{1}{4}$ of circle − half of r^2) and the lune as (semicircle on hypotenuse − white segment).

You can calculate the area of the semicircle on the hypotenuse by using Pythagoras' theorem with semicircles in lieu of squares constructed on the sides of a right-angles triangle

Areas can be surprising! The next task introduces another shape that may be new to you: the arbelos or shoe-maker's knife.

Task 6.1.4 More Lunes and an Arbelos

Figure 6.1e (i) is based on a right-angled triangle. Describe to yourself what you see. Try to work out how to construct it.

Figure 6.1e (ii) uses the vertex of the right angle and the point on the hypotenuse as diameter for yet another circle. Convince yourself, and someone else, that the new circle is necessarily tangential to the hypotenuse.

Which looks larger in each case, the area shaded light grey or the area shaded dark grey?

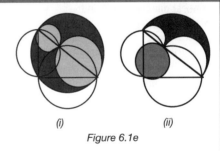

(i)   *(ii)*

Figure 6.1e

Comment

The largest semicircle has the hypotenuse of the triangle as diameter. Two circles have as diameters the other two sides of the triangle and these circles intersect twice, once on the hypotenuse. The two points of intersection are two end points of an altitude of the right-angled triangle. It is possible to use Pythagoras' theorem with semicircles to show that, in fact, the shaded areas are the same area in each case.

This first section was theoretical, working with ideas of infinity and approximation, and is a deliberate contrast to the next section in which it is possible to use paper and scissors, or ATM mats (ATM webref) and Copydex, or Polydron, to make models.

Reflection 6.1

Most of this section is about areas of parts of circles. The underpinning pedagogy used here is to ask you to work on harder problems while the underpinning formula for the area of a circle is still relatively new. Has this been a useful strategy for you?

6.2 LINKING TWO AND THREE DIMENSIONS

The general problem to be investigated in this section is to find the shortest distance between two points on a surface. Many three-dimensional objects can be thought of as made up of a number of two-dimensional faces. This means that the surface of a solid can be formed by a *net* of joined planar regions that are cut out, folded up and glued along edges to make the 'solid' shape. This way of looking at a solid helps with the shortest path question. The first thing to do is to gain familiarity with nets.

Task 6.2.1 Netted

Which of the nets in Figure 6.2a can fold to make a cube?

How many different nets can you find for a cube? For a cuboid?

What shape do you get when you fold up the net in Figure 6.2b? If you make three identical ones, you will find that they fit together.

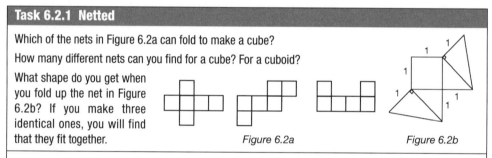

Figure 6.2a   Figure 6.2b

Comment

The principal value in such a task is to develop powers of visualisation, concentration, discernment and mental agility. Did you take time to allow yourself to manipulate the nets mentally to see if they would make a cube?

Notice the difference in effect of the two types of task: 'which of these has … ?' and 'in how many ways can you … ?'. In the first you seed an idea, while in the other you set a challenge. Learners constructing their own nets can challenge each other as to whether a diagram really is a net.

Imagining different nets for surfaces you encounter, such as prisms (including cylinders) and pyramids, exercises your powers to imagine creatively. Note that a net for a cylinder is forced to have a disc attached 'at a point' to a rectangle.

Learners who have not previously encountered nets of surfaces gain a great deal by being challenged to start from a surface and construct a net, and vice versa. Polygonal shapes such as ATM mats (ATM webref), Polydron and so on are ideal for this purpose. Some learners will benefit from working physically, others will benefit from trying to do it mentally. It is through manipulation of familiar objects that people get a sense of relationships and properties that they then can articulate, reaching a point, for example, where they can describe in words how to work out a net for a surface.

One of the many uses of nets is in finding shortest distances on surfaces. As a special case, billiard and snooker players can use an imagined 'net' made of copies of the table to work out how to do banked shots by aiming along straight lines.

Shortest Paths

Imagine a piece of paper with two distinct points A and B marked on it. Now imagine scrunching up the paper in some way. How could you find the shortest path which stays *on the surface* of the paper between A and B?

At first it seems as though you might be able to make use of the scrunching in some way, but that is due to interference between shortest distance through space, and shortest distance staying on the paper. If you flatten the paper out again you do not change the length of the path. Draw the straight line between A and B, and then rescrunch the paper. The line segment still gives the shortest distance staying on the paper, as distinct from moving through space to get from A to B. The next task looks at shortest paths on surfaces.

Task 6.2.2a Shortest Paths

By imagining an ant walking on the cuboids depicted in Figure 6.2c, find the shortest path for the ant to walk from A to B along the surface.

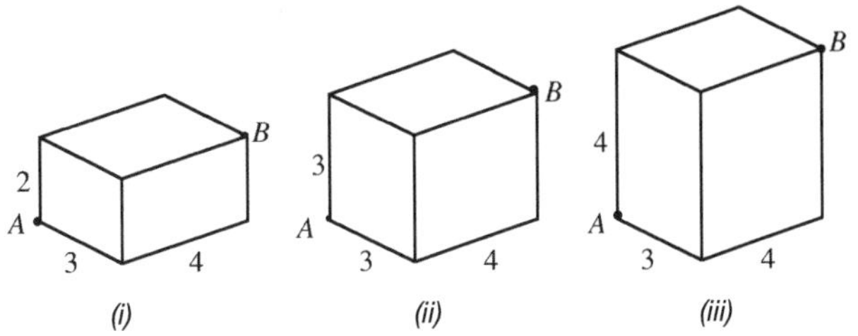

Figure 6.2c

Comment

What did you do in order to imagine an ant taking the shortest possible distance? Did you consider drawing or even making a net? Did you think about tying ribbons around parcels?

If the ant's path is traced on the net, then it must be a straight line between A and B. But there are three possibilities depending on how the net is drawn. Figure 6.2d shows three 'overlapping' partial nets for the cuboid in Figure 6.2c (i), in order to display the three possible routes. It is well worth spending time imagining how Figure 6.2d relates to the cuboid in Figure 6.2c (i), and then using it to imagine the three dashed paths on the surface of the cuboid. You may wish to experiment with the interactive file '6a Shortest Path'.

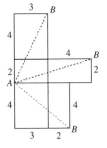

Figure 6.2d

There are other dimensions of possible variation in this task, even when you restrict your attention to cuboids. For example, points A and B might not be at vertices. The variation of just one dimension in the three cuboids of Figure 6.2c might suggest a further task of finding a rule of thumb for deciding which route is the shortest for any cuboid when the ant goes between diagonally opposite corners. Would it be possible for a shortest path on the surface of a cuboid to involve travel on more than three faces?

It seems a short step to move from cuboids to cylinders, since both can be seen as examples of prisms.

A cylinder can be thought of as a prism with a circular cross-section rather than a polygonal one.

Task 6.2.2b Shortest Paths Again

Imagine a cylinder with diameter 4 cm and height 4 cm, with two points marked A and B diametrically opposite each other but at different heights. In your mind's eye, place an ant at point A.

By using a physical net, decide whether it is shorter for an ant to go by way of the top of the cylinder in order to follow the shortest distance from A to B in each of the three cylinders depicted in Figure 6.2e from A to B. Down to what depth from the top on the vertical side is it shorter to stop going over the top of cylinder?

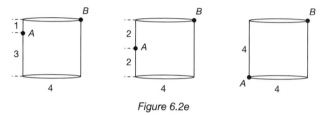

Figure 6.2e

Comment
Actually finding the distance is quite difficult because it requires facts about chord lengths of ellipses. You can make an estimate by approximating the curved journey to a half-circle of radius 2. In fact only for the third cylinder is it shorter to avoid going over the top.

Nets of Cones

By analogy, a cone can be though of as a pyramid with a cross-section that is often (but not exclusively) a circle. A *right circular cone* has its apex vertically over the centre of its base.

Figure 6.2f

Task 6.2.3 Nets of Cones

Imagine the net for the curved surface of the right circular cone shown in Figure 6.2g. Which if any of the nets shown is appropriate? Try to articulate why you think your choice is correct. What curved shape do you imagine would be formed using the other net and gluing the equal sides together?

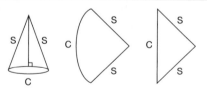

Figure 6.2g

Comment
Most people feel that the sector of the circle 'looks wrong', but in fact it is the one that gives a cone. The edge of the other does not even sit flat on a table when made into a cone! It really helps to manipulate a physical object in order to get a sense of what is happening when the net edges are glued together. Did you think about the distance from the apex of a cone to points on its base?

Using a net makes it easier to think about shortest distances on a cone or the *frustrum* of a cone (the part of a cone lying between the base and a plane parallel to the base; in other words, a cone with its vertex removed to form a face parallel to the base; see Figure 6.2h).

Figure 6.2h

In general, the shortest distance (without using the base of a cone, or the parallel faces of a frustrum) is a familiar curve: it is a part of a helix such as you find on the threads of a bolt, a spiral staircase and a corkscrew.

Shortest Paths on a Sphere

Although a sphere cannot be constructed from a net of planar shapes, it has been necessary for map-makers to try various methods for representing the surface of the earth (which is actually not a sphere but a slightly squashed sphere) on a planar piece of paper. For the moment, using your experience of thinking about co-ordinates in Chapter 5, it is possible to deal with the question of shortest distances that was mentioned briefly there.

Task 6.2.4 Shortest Paths on a Sphere

Given two distinct points A and B on a sphere (see Figure 6.2i), what is the path of shortest distance between them?

Comment
Using an orange or other approximate sphere and a piece of string or elastic may convince you that there is a unique minimum (as long as the points are not diametrically opposite!).

One approach to the task is to imagine a plane through the two given points A and B, and allow it to rotate about the straight line joining A and B. Where the plane cuts the sphere is a path between A and B. The question becomes one of finding the position of the plane that will give a minimum path length on the surface of the sphere. The plane through the centre of the sphere achieves this: the 'great–circle route'.

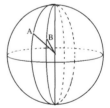

Figure 6.2i

New Polyhedra

Here is a task that demonstrates that great care is needed when 'sticking polyhedra together' to form new ones.

Task 6.2.5 Pyramid Additions

Imagine a cube with six congruent, square-based pyramids glued onto its faces. Describe this new shape. Imagine the apex of each pyramid moving first away from the base of its pyramid, and then towards it. How might your description of the shape change as the apexes vary?

What happens when the pyramids all have their faces inclined at 45° to their base?

Comment

Did you think to sketch a two dimensional version (a square with isosceles triangles sticking out from each side of the square)? Did you think to make some nets (of the final surface, not of the cube and its individual pyramids!) and to fold up the surface to see what it looks like, or were you able to see that sometimes it is 'pointy like a star' and sometimes indented?

The formal name for these shapes is a *tetrakis hexahedron* (meaning roughly, swap each face for four faces, starting with 6 faces), except the 45° case which is a *rhombic dodecahedron*.

Most people find this task pretty difficult without some sort of a diagram, net or model in front of them. Discerning all the details, and holding potential relationships between faces of pyramids is a real challenge. Once you start thinking in terms of properties of the pyramids, and/or using the analogy of isosceles triangles on a square, you can see that at some point pairs of faces are going to merge into a single face. You also find yourself able to think in terms of generalising this construction: take any surface made up of polygonal faces, and stick pyramids on each one to *stellate* the original shape.

No matter how sophisticated you are as a learner, it really helps to find something confidence-inspiring and familiar to manipulate in order to 'get a sense of' the ideas. Sometimes what helps are physical objects; sometimes a familiar example is sufficient. What is familiar and confidence-inspiring to one person may not be to another. However, the key to developing geometric thinking in learners is for them to be encouraged to strengthen their powers to form mental images and to express what they are imagining in words, diagrams and sometimes objects. For example, having made nets for some previously unfamiliar objects, learners can be expected to form mental images for cylinders and cones, by imagining the nets without actually making them.

Reflection 6.2

To think about paths on a surface, it sometimes helps to have the idea of a net, which converts three-dimensional problems into planar problems. What did you notice about the change in thinking as you moved from surfaces to nets and back again?

6.3 VOLUMES OF SOLIDS

This section demonstrates two ways to calculate volumes of solids by making use of other solids whose volumes are already known: volumes by decomposition and volumes by comparison.

Volumes by Decomposition

A tetrahedron is the three-dimensional analogue of a triangle, and the volume of a tetrahedron is (area of base) × (height) /3. The area of a triangle is half the area of the surrounding rectangle. The volume of a prism or cylinder is (area of base) × (height) but it is not obvious why the volume of the tetrahedron (or more generally a pyramid) is one third of this.

One way to seek confirmation would be to use a physical cone and something to fill it, like water, sand or salt. Since this approach would use physical measurements with their attendant errors, it would only be indicative and not fully convincing.

The next task provides a geometrical way to 'see' that the factor of $\frac{1}{3}$ is correct for a square-based pyramid, and is at least consistent with extensions to other types of pyramid.

Task 6.3.1a Volumes of Pyramids

Imagine a cube with base ABCD and with the other vertices EFGH as in Figure 6.3a

Focus your attention so as to discern the shaded part of the cube as the surface of a square-based pyramid. Now imagine another identical pyramid with the front face HGCD as its base, and E as its apex. Finally, imagine the shape of what is left, which is a square based pyramid identical to the other two. Thus, the volume of the original pyramid is $\frac{1}{3}$ of the volume of the cube.

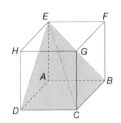

Figure 6.3a

Comment

You could make yourself appropriate nets (use stiff paper or light card and use the net in Figure 6.2b) in order to convince yourself, but you can also do it by systematically picking out the base vertices and the apex, and then checking that it is congruent to the original pyramid. It may help to 'see it' as a rotation of the first pyramid around a horizontal axis.

Draw yourself a cuboid, mark its edge lengths, then try to see the three rectangular-based pyramids corresponding to the three for the cube. Although these pyramids are not congruent, they do have the same volume!

Task 6.3.1b illustrates another approach to the volume of a pyramid.

Task 6.3.1b Volumes of Pyramids again

Imagine the eight vertices of a cube each joined to the centre of the cube (Figure 6.3b).

How many pyramids are there, and how are they related? What then is their volume?

What happens for a cuboid?

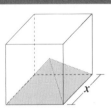

Figure 6.3b

Comment

What is perhaps most important is what you notice about how you had to use your attention, focusing and defocusing, stressing and ignoring, discerning and recognising relationships. That is what learning to think geometrically is really about.

In Task 6.2.1, you worked with the net for a square-based pyramid such as the one illustrated in Figure 6.3a. Three of these pyramids fit together to form a cube. Seeing this mentally is much harder (even when you have them in front of you) than actually doing it physically, but doing it mentally, perhaps after doing it physically, develops facility with mental images, something that many people find difficult until they manipulate, get a sense of and then articulate what they see.

Task 6.3.2 Pyramid Nets

Based on the pyramid net in Task 6.2.1, construct a net for three square-based pyramids that can be folded into a cube.

Comment

Did you think to make three nets and then hinge them together? Which bits did you glue?

Did you try to do it diagrammatically? How might you extend your net to form the net of three rectangular based pyramids that hinge together to make a cuboid? For example, could you do it diagrammatically, having worked on the cuboid version?

Apart from exercising and developing your powers of mental imagery, and your experience of discerning different details, recognising relationships, and perceiving these as properties, the overt aim of these tasks was to suggest that the volume of a cone is the area of the base times the vertical or perpendicular height, divided by three. In the next section a different type of reasoning will be proposed.

Volumes by Comparison: Cavalieri's Principle

Imagine a pack of cards, or a ream of printer paper: you can slide or shear the pile sideways, changing the shape, but not changing the overall volume of the solid that they form. This is because the 'amount of stuff' represented by the very thin cards or paper, is not changed.

Now imagine a picture of such a pile seen from the side elevation, with two possible 'shears' (see Figure 6.3c).

Figure 6.3c

If you imagine the slices to be as thin as you wish, in each case the area is the same. This generalises the idea of the area of a parallelogram which depends only on the base and the perpendicular height (the height of the stack of slices), from which can also be derived the area of a triangle and the area of a trapezium. This is a principle used by Archimedes, and extended by Bonaventura Francesco Cavalieri, an Italian mathematician (1598–1647). His principle is that solids with the same perpendicular height and with the same cross-sectional areas at each distance above the base, have the same volume. More generally, you can calculate the area of a region or the volume of a solid by transforming individual slices so that they form another region or solid whose area or volume you already know.

Task 6.3.3a Cavalieri's Principle

Find the volumes of the solids displayed in Figure 6.3d by relating each to a more familiar solid.

Figure 6.3d

Comment

In the first three cases there are familiar prisms with the same volume. The last are the same but seen from a different position. Notice how you use your mental imagery to alter the shape of a solid by adjusting the relative positions of imagined slices that make it up.

Now it is possible to see the principle in action. The third and fourth examples will need quite a bit of thought.

Task 6.3.3b Cavalieri's Principle Applied

A spherical bead has a hole drilled through it, which turns out to have length $2h$. Find the volume of the material left making up the bead.

Find the area of an annulus by thinking of it as made up of very thin circular strips that can be straightened out to form very thin segments. Use this to find the area of a circle as a special case.

Comment
The volume of the remaining bead is independent of the radius of the original sphere, because the bigger the original ball, the more that gets drilled away to make the remaining hole have length $2h$. Taking slices perpendicular to the hole reveals an area that is the cross-sectional area of a more familiar object.

Task 6.3.4 Volume of a Spherical Ball

Imagine a cylinder with the same radius as a ball, and the same height as the ball (equal to the diameter of the ball). Now imagine a cone with its vertex at the centre of the cylinder and its base the base of the cylinder. Imagine also another copy of that cone upside down so that its 'base' is the top of the cylinder. (Figure 6.3e may help your imagination!)

Calculate the area of a slice of the sphere and a corresponding slice of the region between the cylinder and the cones.

Comment
It might help to draw your own version of the diagram.

Figure 6.3e shows a slice of the sphere (radius r) and of the region between the cylinder and the cones, at a height of y above the middle of the cylinder.

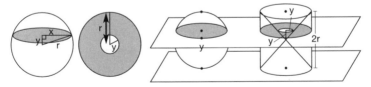

Figure 6.3e

The cross-section of the sphere is a circle. If the radius of this slice is x, then $x^2 = r^2 - y^2$ (Pythagoras' theorem). The cross-section of the sphere has area πx^2, which equals $\pi r^2 - \pi y^2$. But this can be thought of as the area between two circles of radii r and y respectively.

The cross-section of the cone through the vertex is made up of two right-angled triangles of height r and base r, in other words two 45° right-angled triangles. Therefore the angle formed at the centre of the cone is 90°.

So, the radius of the cone at a distance y above the centre is also y, and the area of the shaded cross-section is $\pi r^2 - \pi y^2$.

At every level above the base the sphere has the same cross-sectional area as the cylinder minus two cones, so by Cavalieri's principle their volumes must be equal. That is:

$$V(\text{sphere}) = V(\text{cylinder}) - V(\text{two cones})$$
$$= \pi r^2 (2r) - 2[\tfrac{1}{3}\pi r^2(r)]$$
$$= 2\pi r^3 - \tfrac{2}{3}\pi r^3$$
$$= \tfrac{4}{3}\pi r^3.$$

The surface area of a sphere (ball) can be found by switching approaches and thinking of the ball as made up of very many little pyramids, each with their apex at the centre, and their bases approximately filling out the surface. Adding up all their volumes will give the volume of the ball, which means that

$$\tfrac{4}{3}\pi r^3 = \tfrac{1}{3}rb_1 + \tfrac{1}{3}rb_2 + \tfrac{1}{3}rb_3 + \tfrac{1}{3}rb_4 + \tfrac{1}{3}rb_5 + \ldots$$

where b_k denotes the area of the base of the kth pyramid. Since the bases approximate the surface of the sphere, the surface area A must be related to the volume by

$$\tfrac{4}{3}\pi r^3 = \tfrac{1}{3}rA$$

So surface area = $4\pi r^2$.

The surface area of a sphere looks familiar if you have thought about surface area of cylinders: the surface area of a sphere is the same as the surface area of the cylinder of the same radius and height (ignore the ends), as Archimedes noticed. (See Figure 6.3f.)

Figure 6.3f

Here is another application involving shapes of containers.

Task 6.3.5 Glasses

Figure 6.3g shows four glasses (not drawn to scale).

The second and fourth are cylinders with a hemispherical bottom. Find the vertical height of the third and the diameter of the fourth. Then put them in order from largest capacity to smallest by estimation. Check your conjecture by calculating the volumes.

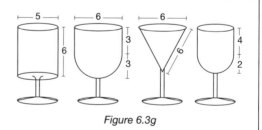

Figure 6.3g

Now estimate the wine levels if each glass was half full. Check your estimates by calculating the levels.

Comment

While you are likely to have used your prior experience in making your estimations, you may still have been surprised when you checked your answers. Learners in school have fewer prior experiences to draw upon and their estimates and reasoning usually range very widely. An estimate in such examples is not intended to be a guess, although pupils often think that this all that is required; anyone making an estimate should be able to support their decision with some justification. Now reflect on your justifications and, if they did not produce the correct orderings, see if you can see what went wrong.

A handy conversation starter at mealtime is to ask people which is greater, the circumference or the height of their glass. Apart from some champagne flutes, people are usually astonished by what they discover when they make a comparison.

Reflection 6.3

Tangrams are often used to work on visualisation in two dimensions as well as exposing learners to the invariance of area when pieces are cut up and moved around. What have you learned about visualising three-dimensional objects and cutting them up?

6.4 SPHERICAL GEOMETRY

This section is a brief introduction to another geometry that makes use of circles, but in somewhat unusual, yet nevertheless powerful, ways. One of the aims of the tasks in this section is to present some less familiar geometrical thinking.

Although geometry is traditionally about points, lines and then circles and other shapes, mathematicians discovered that the objects called 'points' and the objects called 'lines' need not actually be what are familiarly known as points and lines. In other words, what emerged was that geometry is about the study of relationships that are treated as properties, and not about specific objects as such. Spherical geometry provides an opportunity to work with familiar properties without relying on familiarity with the objects themselves.

When you are only concerned with moving a few hundred kilometres, you can assume you are moving on a flat surface. Navigators trying to sail to distant continents on the surface of the earth needed to be able to do calculations that took into account the spherical surface of the earth. The developments they made turned into an important geometry on the sphere.

Take as points, the points on a sphere (not a ball, just the surface). For clarity, call them S-points for surface points. Now consider what the term 'line' means in this case. Call it an S-line. An 'S-line' between two S-points must involve only other S-points. Familiarity with great circles and distances on a sphere suggest that an S-line is the great circle through the two points. What properties of 'lines' are also possessed by S-lines?

Task 6.4.1 S-Line Properties

On a plane, lines have the following properties. Which of these are also possessed by S-lines?

On a plane, when three distinct points are collinear, one is between the other two. Does this work for three S-points on an S-line?

On a plane, there are infinitely many lines through a given point. What about S-lines through an S-point?

On a plane, there is a unique line through two distinct points. What happens on the sphere? (S-points that are diametrically opposite might be problematic.)

On a plane, two lines meet in a unique point. What happens on a sphere?

On a plane, through any point not on a given line, there is a unique line through the point and perpendicular to the given line. What might perpendicular mean for S-lines? Are there any special cases?

On a plane, through any point not on a given line there is a unique line through the point and parallel to the given line. What might 'parallel' mean on a sphere?

Comment

There are some similarities, but some differences. These differences become accentuated when triangles are considered.

The notion of an angle between two S-lines can be based on ordinary planar angles as follows. S-lines, being great circles, are formed by a plane through the S-line and through the centre of the sphere. Now imagine a plane tangent to the sphere at an S-point where two S-lines meet. This plane is cut by the great-circle planes and so an angle is formed on the tangent plane (see Figure 6.4a). This serves to measure the 'angle between two intersecting S-lines.

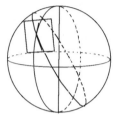

There is no concept of parallel for S-lines, because every pair of distinct S-lines intersects in two diametrically opposite S-points.

Note that lines of latitude are *NOT* S-lines (with the exception of the equator).

Figure 6.4a

Without a notion of parallelism, shapes such as square, rectangle and parallelogram have to be defined in other terms, such as angles. Start with triangles.

To form a triangle, three S-points are needed, with the S-lines that join them. Imagine an S-triangle formed by the north pole (N), two S-lines, and points where the S-lines meet the equator. There are two angles of 90° and 90° where the S-lines meet the equator, and an angle that can vary between 0 and 180° at N. This raises serious concerns about the sum of the angles of a triangle. In fact, because of the absence of parallelism, the Euclidean proofs about sums of angles in a triangle do not work, and indeed the sum of the angles of an S-triangle is always greater than 180°.

Task 6.4.2 S-Quadrilaterals

What can be said about the sum of the angles of an S-quadrilateral?

Is it possible to have an S-quadrilateral with four right angles?

Is it possible to have an S-kite?

Comment

Did you find yourself reaching for a ball, or ball-like object on which to draw? If so, were you able then to put it to one side and imagine what you drew? Can you sketch it? Finding 'the answer' is much less important that extending your powers to imagine and to express what you imagine.

Task 6.4.3 S-Circles

What would S-circles look like? How might these relate to lines of latitude relative to some choice of north pole?

Comment

Any S-point on the sphere can be taken as the north pole, so all S-circles are indeed lines of latitude relative to some choice of pole.

Istvan Lenart (1996) has developed workbooks and models for exploring spherical geometry in detail. For some pictures of shapes in spherical geometry, one particularly interesting website is Mathworld (webref).

Reflection 6.4

What did you find you had to do in order to think about a less familiar 'geometry'?

6.5 PEDAGOGIC PERSPECTIVES

One of the inner aims of the tasks in this chapter, particularly in sections 6.3 and 6.4, was to present some less familiar forms of geometrical thinking, so that you would be reminded of what it might be like for learners. Some ideas are familiar, some less so, and some familiar ideas develop in unfamiliar ways. Many of the feelings you had as you worked through the tasks will be similar to those experienced by other learners working through new material.

Some tasks explicitly invited you to use familiar objects (nets) in order to get a sense of an idea (shortest distance on surfaces), while other tasks deliberately tried to get you to make use of less concrete but equally familiar ideas to manipulate as diagrams or even verbally or symbolically. Tasks that provide familiar objects to manipulate but which do not call upon learners to 'get a sense of something' and to try to bring that to articulation, are likely to be less successful than tasks that do call upon learners' powers. Getting learners 'doing' something is relatively straightforward; getting them to do something that sparks the need to talk about it, and to refine that articulation so that it can be captured in diagrams, words and symbols, requires tasks that have some challenge and ambiguity, but also requires establishing ways of working in the classroom that promote transitions from 'getting a sense of' to 'articulating', through discussing with each other before being pushed to written records. These ideas are taken up in Chapter 14 more explicitly under the headings of Manipulating–Getting-a-sense-of–Articulating (MGA) and Do–Talk–Record (DTR). The tasks have also called upon you to discern details, to seek out relationships, to perceive properties and to reason on the basis of those properties.

All these ways of thinking about pedagogy are based on the assumption that you as a learner need to become familiar with the shapes or ideas that you are dealing with first, which might involve looking at them, handling them and drawing them from different perspectives. Only after some continued exposure are you likely to get a sense of them, to be in a position to recognise relationships and then to perceive these relationships as properties independent of the particular situation. As you begin to articulate your ideas you start to reason about the properties, informally at first and then with more rigour.

In saying that, it is not intended to imply learning happens in a steady linear pattern of development. Learners move back and forth through different forms of attention, through different 'levels', often very rapidly. Development is not so much 'linear' as more like a spiral, in which learners return to previous ideas but with greater sophistication. Susan Pirie and Tom Kieren (1994) describe these movements eloquently as a process of 'moving out and folding back' in their 'onion' model of understanding. However it is described, learning takes place over time, as a maturation process, not as a series of steps on a staircase of learning. Time is needed to develop familiarity with new concepts before learners can look at them analytically, but often the pressures of the classroom mean that not enough time is given to the need for establishing familiarity and confidence.

It is also possible to overdo the confidence-building aspects by assuming that learners know less, and have access to fewer powers, than they in fact do. For example, some people think that children are more familiar with three-dimensional objects than two-dimensional ones and that geometry can more effectively start with solids rather than plane shapes. If you think about the geometrical objects young learners are given to manipulate, they are all actually three-dimensional and usually solid. From the start learners have to work out what features are being stressed by the teacher in the way things are talked about, and what features are to be ignored (thickness, colour, material, …). This does not mean that solid geometry should be taught before plane geometry, but that they both might be taught in a more integrated way.

Real-world applications are only of any use if learners can appreciate them on the basis of their experience. Where this is possible, early use of applications may provide concrete scaffolding for abstract ideas.

Young children are adept at picking up new words. For new terms to become part of their own vocabulary, it is essential that learners are called upon to express ideas and conjectures using the new terms. By being aware of conflicts in use of technical terms in ordinary language, and by being alert to ambiguities in what and how learners express themselves, it is possible for teachers to circumvent some possible misuse or misconceptions, or to work with learners at clarifying issues. This means seeking tasks that challenge learners to do things, to express their thinking to each other and to the teacher, in a conjecturing atmosphere in which everything is said in a spirit of intending to modify it if needs be.

Reflection 6.5

What did you have to do in order to think about unfamiliar 'geometries'?

7 Geometrical Reasoning

This chapter is about the nature and role of reasoning in geometry. The purpose of reasoning is to justify conjectures, to prove that conjectured facts are indeed always facts. Sometimes a proof is convincing locally (you can follow all the steps but the overall picture or the structural 'reasons' remain mysterious); sometimes a proof is explanatory (the reasoning shows why something must always be the case).

7.1 THINKING ABOUT PROOF

I put before you my two main conjectures:

What is important about geometry is being aware of the fact that there are facts, rather than mastery of some particular few facts.

Geometry takes place in a world of forms and images, entry to which is gained through the power of mental imagery, augmented and extended by dynamic images, drawings on paper and discussion with colleagues.

(Mason, 1989)

As mentioned earlier, geometry was not part of the ancient Greek curriculum. It was included in the medieval quadrivium (fourfold curriculum) and in Victorian times a two-column layout of assertions and justifications became a central part of UK grammar school education. Geometry was credited with teaching young people how to reason effectively and to educate their awareness about spatial relationships.

In the 20th century, geometry and its reasoning faded from the UK mathematics curriculum, largely because it seemed very difficult to teach learners to prove things, so learners were encouraged to memorise proofs instead. Recent curriculum development in England has recognised a place for geometry: there are geometrical facts that govern how the material world works.

More recent research in mathematics education has revealed some of the reasons why learners find proof challenging. The van Hiele (1986) levels give teachers a perspective on adolescents' experience of moving from truth and rules belonging to adult authority to discovering that mathematical truth comes from the mathematics itself and so is in the purview of learners themselves. Thus reasoning and proof in geometry offer learners an opportunity to become their own authority, within the social conventions of how mathematical reasoning proceeds, so that they need not take the word of any authority other than reasoning and mathematical structure.

Most of the difficulties encountered with proof have to do with knowing what you are allowed to assume, and what you are expected to justify. To prove, you need to appreciate that properties can apply to a variety of objects, and that reasoning proceeds on the basis of established and accepted properties only.

Being aware of how this way of working developed is sometimes helpful to both teachers and pupils. To sketch in a little historical background:

The father of all the Western philosopher mathematicians is claimed by many to be Pythagoras (sixth century BCE) who was described by the historian E. Bell (1953, p. 20) as 'mystic, mathematician, ... one-tenth of him genius, nine-tenths sheer fudge'. Bell suggests that Pythagoras was the first European to realise (or to learn from those with whom he came into contact from other cultures) that proof must proceed from agreed properties and explicit assumptions.

In the 200 years following Pythagoras' teaching, there was continuing, but sporadic, geometrical development in several Greek city states. It became increasingly understood that from a set of arbitrary postulates new ideas could be reached by close deductive reasoning. Proof Four in section 7.2 is an example of organising such a chain of deductive reasoning.

Socrates (469–399 BCE) and his pupil Plato (427–347 BCE) both worked in Athens. Euclid (330–275 BCE) worked in Alexandria, a new city at the mouth of the Nile founded by Alexander and ruled by one of his generals, Ptolemy. Euclid collected geometrical work and codified it into 13 books. The first six books are about plane geometry; books 7–9 deal with number theory; book 10 is about Eudoxus' theory of irrational numbers and books 11–13 concern solid geometry, ending with a discussion of the properties of the five regular classical polyhedra and a proof that there can be no more than these five. Examples of all five polyhedra can be found in museums in Edinburgh and elsewhere, carved in granite balls by Neolithic farmers.

Euclid was a little older than Archimedes but two generations younger than Plato. A contemporary of Euclid was Aristarchos of Samos (who decided that the earth went round the sun and not, as previously thought, the other way about). For more information, see St Andrews (webref).

What Euclid offers to us in books 1–6 is a logical structure for plane geometry developed from a few postulates, or axioms. The axioms are not unique; a different set to the ones proposed by Euclid could be used. Leading up to the time of Euclid, the Greek geometry programme insisted that all geometry had to be made with a straight edge and pair of compasses only. From these two tools, the world of classical geometry was open. It was also a model of 'postulates asserted, theorems deduced'.

Ideas were communicated in writing between the various 'Schools' (groups of intellectuals who necessarily were landowners and so independently wealthy). What we have is written evidence that these ideas were communicated around the Mediterranean but we cannot know from the written evidence how much thought experiment and visualisation were used by the founders of Euclidean geometry in their creative teaching and thinking.

Fragments of stories have survived about individuals working as teachers. Plato comes over as a superb teacher. Euclid does not come over as a sensitive teacher. He is reported to have dealt with Ptolemy in the following way: Ptolemy attending one of Euclid's seminars asked for a simpler chain of reasoning. Euclid's cutting response was 'There is no royal road to geometry'. On another occasion, a member of the Academy asked 'What's the use of this (theorem)?' Euclid's response was to address the rest of the group and say 'He wants to profit from his learning, give the man a penny'.

A problem in working with 'geometrical proof' at school or university level is that learners find it hard to enter into the rules of the game. What are the givens? What do we need to prove or deduce? So often the 'fact to be proved' appears to be obvious. Why are we taking all this effort to 'prove' when I know it is the case? Why won't a few examples suffice, or perhaps a diagram in dynamic geometry software with the opportunity to 'drag' in order to test for generality?

Historical records identify only what authors chose to articulate and record: their best thoughts. To try to appreciate what they were thinking and how they reached their conclusions it is necessary to reconstruct objects that might have been manipulated, calculations that might have been done, special cases that might have been tried. To teach geometrical thinking requires tasks for learners that engage them in appropriate manipulations, which give them opportunities to get a sense of relationships and to see these as properties independent of the particular situation, and finally to reason solely on the basis of agreed properties.

This chapter takes the idea of chains of logical reasoning and builds a number of 'proofs' from different starting points. The task, each time, is to ask yourself:

- Do I know what the givens (assumptions) are? What properties are used in the reasoning?
- Is there enough manipulating to get-a-sense-of for me to believe this proof?
- What needs to be added, or telescoped, if I were to use this proof with pupils?

Reflection 7.1

To what extent are you aware of learners around you with preferences for spatial thinking? What is their entitlement to geometric as opposed to symbolic approaches to reasoning?

7.2 PYTHAGORAS' THEOREM

Probably the most famous of mathematical theorems, it is thought by scholars that Pythagoras might have learned the theorem, or at least the conjecture, from scholars he met in Babylon, some of whom could have come from China, as there is a long history of the result in Chinese literature. Specific examples are attributed to a Chinese manuscript *Chou pei suan ching* (*c.* 1100 BCE) and Babylonian tablets (dated variously between 1800 and 1000 BCE). See Swetz and Kao (1977). Euclid may have been the first to state and prove both the theorem and its converse explicitly. In this section, several types of reasoning will be offered for the theorem. Over 350 different proofs are known. (The converse states that if the area of the squares on two sides of a triangle equals the area of the square a the third side, then the triangle contains a right angle.)

Proofs of Pythagoras' Theorem

The first two tasks approach a proof of Pythagoras' theorem from the point of view of tiling.

Task 7.2.1 Tiling a Floor

Imagine you are going to tile your bathroom with a pattern that repeats a finite block of tiles infinitely (if the floor were big enough!). You have square tiles in two sizes and want to use both.

How many different ways can you find to tile the floor using both large and small tiles? What happens if you restrict yourself to using the same number of large and small tiles in the core repeating pattern block?

Does it matter what the two sizes are?

Comment

Were you able to visualise a pattern or patterns in your head? Would you prefer to do work practically with gummed or card squares or squared paper, or ATM mats (webref)?

How do you ensure the pattern repeats? How are you sure that you have a repeating pattern that extends infinitely?

If you can catch the moment when you realised that your pattern would extend indefinitely, if you can catch yourself organising how the tiles will fit together and the moment when you realised that 'it would always work', you have a taste of what it feels like to (find a) reason. Working with cardboard squares you may have found yourself arranging and rearranging; working with drawn diagrams you may have found yourself erasing or starting again several times.

Some patterns may depend on a particular relationship between the sizes of the two tiles. There are three different types of solutions that we will discuss in turn below, with commentary.

Type 1: An Arrangement of Rows or Columns of All the Same Type

Note the possible variations suggested by the double-width row of small squares in Figure 7.2a. Multiple rows of the same sized tiles could be displaced slightly amongst themselves as well. For the purposes of proving Pythagoras' theorem, all these variations are considered to be of the same type.

Note that the relative sizes of the two tiles does not matter in general, but that if equal numbers of small and large tiles are required, this type of patterning will need to be excluded.

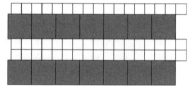

Figure 7.2a

Type 2: Blocks of One Size Tile Bordered by Infinite Rows and Columns of the Other

Note the dimensions of possible variation that can be indicated through just the example shown in Figure 7.2b. For the purposes of proving Pythagoras' theorem, all these will be considered to be 'of the same type'.

Note that the relationship between the sizes of the tiles determines to some extent the possible sizes of blocks of the large tiles: you need to know the lowest common multiple (LCM) of the two tile

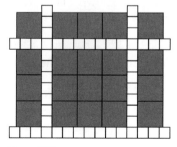

Figure 7.2b

edge lengths in order to know what the possibilities are. Note, too, that you can have a repeating pattern that makes use of a large core pattern: there could be blocks of large tiles of different multiples of three, interlaced with the rows and columns of small tiles, and the whole thing repeated over and over. Restricting attention to tilings using the same number of large and small tiles within the repeating core pattern excludes this kind of tiling as well.

Type 3: Arrangements of Running Diagonals with the Small Tiles as Spacers

Tilings of type 3, illustrated in Figure 7.2c, might not have occurred to you immediately, though once you try to break out of having rows and columns, it arises quite naturally. Many Greek and Roman mosaic-makers used this third solution. It appears in kitchen or bathroom tiling designs today. Thinking about dimensions of possible variation leads to using more than one small tile (depending on the relative sizes).

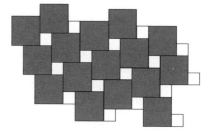

Note that each of these tilings has a repeating pattern in any given direction, and also rotational symmetry. If the core pattern has to use the same number of large and small tiles, there is just one tiling (with different orientations).

The fact that this third tiling exists at all is due to Pythagoras' theorem! At first glance, this statement appears to be out of line with the space-filling task you have just done. In

Figure 7.2c

constructing the tessellation you were likely to have been working to be systematic in placing the small squares around a large one and then moving to place the next large one, and so on. Questions such as: 'Am I sure that this pattern will continue to infinity?' will probably not initially have come to mind. Once you discern the repeating block of a large and a small tile attached together (note that there are four ways of seeing this), the fact that the tiling repeats indefinitely becomes obvious.

Pythagoras' theorem states a connection between the lengths of the sides in a right-angled triangle. But, because the theorem connects the squares of the lengths of the sides, it is, in effect, a theorem about area, and that is what connects it to the tiling.

Task 7.2.2 From Tiling to Pythagoras' Theorem

Look at Figure 7.2d. Focus on and mentally stress first the two-sized tiling of the plane. Then stress the tiling by larger tiles that are tilted.

Now, seeing the tilted tiles as derived from the two-sized tiling, convince yourself that the tilted tiles are in fact squares. That they may look like squares is at best suggestive but not necessarily convincing. All you can be sure of is the fact that the two smaller tiles are squares, and that some of the triangles are similar.

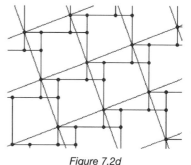

Gaze at one of the tilted squares and see it as made up of pieces of the tiling; rearrange the pieces mentally to discover the area of the tilted square in terms of the areas of the tiles.

Figure 7.2d

Convince yourself that the relationships you detect in one particular case are actually properties (the whole relationship extends to infinity).

Comment
Your power to imagine is being called upon explicitly in this task. This power can be developed with practice. Practice, in this case, means working on the detail of what you see, including relationships amongst features that you discern in the diagram. A way to work on this is to 'say what you see' either out loud, or to yourself.

If you wish to explore this image interactively, to help you develop your power to imagine, open the interactive file '**7a Tiling Pythagoras**' on the CD-ROM.

The next task is a development of the previous one. Use either the interactive file '**7a Tiling Pythagoras**' or make a copy on tracing paper of slanting squares that fit over the drawing you made for Task 7.2.2.

Task 7.2.3a Another and Another and More

Convince yourself that, in *any position*, the area of a 'tilted' square will be equal to the area of one small square plus one large square.

Comment
The fact that translating (sliding) the large squares does not change their area, and so leaves the area relationship is easier to see in general than to see how the areas add up in some particular cases. For example, see Figure 7.2e.

Figure 7.2.e

Is this set of pictures a 'proof' of Pythagoras' theorem? This is a hard question, because it depends on the criteria for proof; these criteria have changed over the centuries as mathematicians have found more flaws in proofs that were previously accepted, and become more demanding as a result. A more helpful place to start is with a simpler question: 'Am I convinced by my visualisation work on the areas of the squares that Pythagoras' theorem is true?'

It is useful to think of three states in being convinced:

- convincing yourself
- convincing a friend
- convincing a sceptical outsider.

Convincing a friend who nevertheless asks questions in order to understand what you are saying can help you to organise your thoughts and to become aware of places where someone who has not already gone through your thought processes may have questions or objections.

The purpose of convincing a sceptic is to produce reasons that will convince someone even if you are not there to answer any objections they may raise.

The reason for working on multiple 'proofs' of a result such as Pythagoras' theorem is to experience different ways in which a proof can illuminate underlying structure. Throughout this chapter, you will be presented with more than one 'proof' of a number of theorems. Each time you work through the various 'proofs', your task is to work at the structure of the 'proof' using the three levels above.

Task 7.2.3b Discerning Pythagoras' Theorem – Proof One

Find in Figure 7.2d a version of the familiar 'three squares round a right-angled triangle'.

Comment
For some people, especially those already familiar with Pythagoras' theorem, finding the three squares round a triangle is an 'aha!' moment, a moment of insight, where the connection between tiling and Pythagoras' theorem is established.

It is not sufficient to put faith in a long line of mathematicians who have proclaimed this or any other theorem. Mathematics provides a domain in which the authority for what is true or false shifts from 'respected others' to individual mathematical reasoning. Each learner can become the authority for why something is the case; it is not necessary to accept someone else's say-so. Learning to become a sceptic, to challenge other people's reasoning, is an integral part of learning to reason for yourself: it means asking 'why?' and 'what is the reason for …?' in place of 'show me an example' and not accepting 'if I say so' or 'it just is' as sufficient reason.

The word *theorem* is from the Greek meaning 'a seeing': a theorem is a 'seeing' of relationships; a 'proof' is a way of seeing how those relationships are due to properties that have been previously agreed and accepted. So looking at Figure 7.2.d, you need to convince yourself, by gazing at the figure, and discussing inside your head, why the area of the slanting square is the same as the sum of the areas of one big square and one small square. Then you can use this awareness to direct the attention of a friend to seeing what you see and, finally, you can articulate this to a sceptic.

Discerning details in geometrical diagrams can be like looking for something in a cupboard: sometimes it is right in front of you but still you do not see it. The fact that figures look as though they have a particular property does not mean that they always do, in all cases: reasons are needed, not appearances. This is why some generations of mathematicians eschewed diagrams altogether, because they are particular cases and can mislead. But they can also show the way to a general and convincing proof, especially for those learners with a preference for, or strength in, spatial thinking.

The next task introduces 'Proof Two'. This proof is very different in nature, at least on first sight. Take a right-angled triangle, with hypotenuse c, and rotate it round four times to form a square, with a square hole in the middle. This diagram, put together with the injunction 'behold!', is thought to be due to Bhaskara, an Indian mathematician of the twelfth century (who also wrote the first book to use the decimal number system systematically).

Task 7.2.4 Behold! Pythagoras' Theorem – Proof Two

Look carefully at Figure 7.2f.

Convince yourself, and prepare yourself to convince a friend and then a sceptic, that the 'hole' is indeed a square.

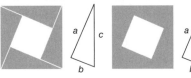

Figure 7.2.f

By calculating and comparing areas of the large square, four copies of the triangle, and the hole, convince yourself that Pythagoras' theorem always holds.

Comment

The reasoning implied by 'behold!' is very different from the reasoning in Proof One. Here, you are asked to calculate and as a result your thinking is directed towards being much more algebraic, relating measurements of edges to measurements of area, rather than looking at area directly. To appreciate the generality, you have to see through the particular diagram and 'see' that neither the construction (placing four copies round a hole), nor the algebraic reasoning depends on the sizes of the sides of the triangle, which can be any positive lengths and not just integers.

Proof Two used the fact that each side is labelled with a letter. The symbols (letters) represent lengths of the sides of the triangle. The rest of the proof manipulates the letters to achieve the required result. It is possible to work with the symbols and 'go into automatic processing mode', ignoring their meaning and only asking what the result means at the end of a series of manipulations. This is both a source of power and also a source of problems for many learners struggling with algebra. Many who can work visually within geometry are completely turned off by a symbolic proof. This raises then the question of what the mode of proof offers in terms of convincing different people with different ways of thinking.

The third proof is based on area rather than algebra.

Task 7.2.5 Pentagons: Pythagoras' Theorem – Proof Three

Use Figure 7.2g to convince yourself, then a friend, then a sceptic that the dark grey pentagon is equal in area to the light grey pentagon. Then restrict your attention to prove Pythagoras' theorem. Where is the generality?

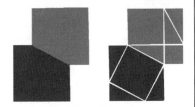

Figure 7.2g

Comment

The light and dark grey pentagons are *not* reflections of each other. They *are* congruent, because one can be rotated onto the other about the mid-point of the segment they share.

Were you able to work solely with the right-hand drawing by ignoring the white division lines at first, or was it necessary to use the left-hand diagram at first?

In order to generalise the result, it is vital to see how the construction works for *any* right-angled triangle, not just this particular one.

The next proof provides all the assertions required, but scrambles the order. You are invited to unscramble the assertions.

Task 7.2.6 Step-by-Step: Pythagoras' Theorem – Proof Four

Accompanying Figure 7.2h is a set of statements, referring to the lengths and triangles from the figure. The statements are set out in random order. Make a copy of these statements, cut them out, and re-arrange them to give a logical step-by-step proof of Pythagoras' theorem.

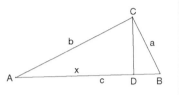

Figure 7.2h

You may feel that you need to insert more statements so that your proof flows with no sudden gaps. Use blank statement cards as necessary.

$\triangle ABC$ has a right angle at C	$\angle DBC$ is common
$\dfrac{AC}{AB} = \dfrac{AD}{AC}$	$a^2 = c^2 - b^2$
$\dfrac{c-x}{a} = \dfrac{a}{c}$	$\angle ACB = \angle ADC = 90°$
$\triangle ACD$ and $\triangle ABC$ are similar	Let $AD = x$
$a^2 = c(c - x)$	$\angle DAC$ is common
$\dfrac{b}{c} = \dfrac{x}{b}$	Drop the perpendicular from C to AB, meeting AB at D
$\dfrac{BD}{BC} = \dfrac{BC}{BA}$	$b^2 = cx$
$a^2 = c^2 - cx$	$\triangle BDC$ and BCA are similar
$\angle BDC = \angle BCA = 90°$	$DB = (c - x)$

Note: when two triangles are written next to one another with the statement 'are similar', the letters have been arranged so as to correspond.

When you are satisfied that you have constructed a logical chain of reasoning, present it to somebody else and check if they can follow the steps you have set out. Allow them time to rearrange the statements or to introduce new ones that you perhaps considered to be superfluous.

Comment
It is quite likely that another person will produce a different ordering of these statements. It is also likely that both their order and your order will be valid logically. Reflecting on this can be productive, because it lies at the heart of a daily dilemma for mathematics teachers. You arrive to look over the shoulder of a pupil working on a problem that you know. The steps set out on the pupil's page are not the steps that you took when you solved that problem. How do you enter the logical chain that the pupil has started to construct and, from a standing start, offer constructive ways of moving on from their 'being stuck in the moment' position?

Exploiting the Different Proofs

Pythagoras' theorem states that 'In a right-angled triangle … '. Where has the fact of a right angle been used? Is it necessary?

Looking back at the basis of Proof Two, the two images in Figure 7.2i make it quite clear that if the triangle is not right-angled, then the hole in the middle is no longer a square and so the proof breaks down. In this visual proof the relationship only holds if the triangle is a right-angled triangle. There is an interactive version of the diagram '**7b Pythagoras without a right-angle**' on the CD-ROM.

Figure 7.2i

It is worth articulating to yourself what happens when the angle is larger than, or smaller than, 90°. Every theorem in mathematics hides a surprise (otherwise it would not be worth stating much less proving!). Here it is amazing that it is just 90° that yields the simple relationship between the areas and hence the sides of the triangle. It takes a bit of drawing and thinking to see why Proof Three using pentagons breaks down when the angle is not 90°.

Proof One holds more surprises: the squares used in Proof One can all be deformed into parallelograms as in Figure 7.2j.

Unlike the other proofs, the basic diagram still seems to work! One triangle has been shaded in. Its edges can be seen to

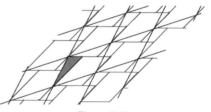

Figure 7.2j

determine the edges of the parallelograms. By gazing at the diagram, you may be able to convince yourself (and then a sceptic?) that the large parallelograms are equal in area to the sum of one each of the other two sizes of parallelograms.

This result seems to be due to Pappus of Alexandria, who lived and worked in Alexandria 600 years after Euclid (900 years after Pythagoras) and who produced a number of important theorems that still bear his name. Pappus' area theorem gives a way of constructing a parallelogram on the third side of any triangle in such a way that its area will be the sum of the areas of two parallelograms constructed on the two smaller sides. It is a beautiful and astonishing result.

Proof Four not only justifies Pythagoras' theorem but also yields an important generalisation, known as the law of cosines. Consider Figure 7.2k, (known as Vecten's diagram) which is a development of the diagram you were invited to construct in Task 5.4.2. Three altitudes have been added to the diagram and these have each been extended to cut a square into two rectangles. You saw in Task 5.4.2 that the areas of the four triangles are equal in magnitude.

Figure 7.2k

In addition, *each pair of rectangles with the same density of grey has the same area.*

One way to prove this result proceeds by finding that the area of each one of the pair is the product of the cosine of the angle at their shared vertex, and the sides of their respective squares. In other words, the area of each grey rectangle is of the form $ab \times \cos C$, where C is the angle between the two sides a and b.

The area of each square is the sum of the other two squares minus the two extra, equal-shaded rectangles. Thus $c^2 = a^2 + b^2 - 2ab \times \cos C$ and this is the law of cosines that is the generalisation of Pythagoras' theorem. In the case of a right-angled triangle, $\cos 90°$ is zero.

In order to appreciate the full scope of generality, it is also necessary to consider what happens to the diagram when one of the angles in the triangle is obtuse.

For more on Pythagoras and the theorem that bears his name see the website St Andrews (webref) about history of mathematics and the website 'Cut-the-Knot' (webref). (See References at the back of the book for web addresses.)

Reflection 7.2

What differences have you noticed in your degree of conviction and in what you had to do to understand the four proofs offered? Which ones offer the most opportunity for you to evoke surprise in learners with whom you work?

What changes are you aware of as you move from discerning specific details, to recognising relationships, to perceiving properties that hold in general, to reasoning on the basis of properties?

7.3 REASONING ABOUT IN-CIRCLES AND EX-CIRCLES

Pythagoras' theorem is just the tip of the iceberg, albeit an important tip. There are thousands of surprising relationships that can be established in geometry. This section considers some circles connected with triangles.

The section involves thinking about the in-circle and the ex-circle of a triangle. The in-circle of a triangle is a circle drawn inside the triangle in such a way that it is tangential to all three sides. The ex-circle is a circle tangent to all three sides of a triangle, but two of those points are on sides extended beyond the triangle. The in-circle and one of the three ex-circles of a triangle are shown in Figure 7.3a.

Three proofs of a formula for the radius of the in-circle are presented, in this section, followed by an opportunity to revisit the same ideas with ex-circles. One outcome is a formula for the area of a triangle known as *Heron's formula.*

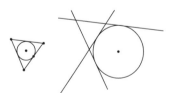

Figure 7.3a

In-circles

As an introduction to in-circles, you are invited to start by imagining.

Task 7.3.1 Constructing the In-circle

Ask someone to read these instructions out to you, with pauses in between, so you are able to do what is suggested. Do not resort to paper and pencil until you really cannot take on board any further instructions.

Imagine a blank flat sheet. On this sheet make a triangle. Move your triangle around so that it is not a special triangle, not equilateral, not isosceles. When you are sure that you have a general triangle label the vertices A, B, C.

Imagine that you want to draw a circle that touches AB and AC. In your head construct the bisector of the angle at A. Place a point on this bisector and draw the circle that just touches the two sides AB and AC. Pull the point backwards and forwards along the bisector line watching the circle growing and shrinking so that it always just touches AB and AC.

Now draw the bisector of angle B. Construct a new circle with its centre on that angle bisector. Pull the circle in and out. Pull it gently out until its centre is at the point where the two bisectors meet.

Open your eyes and provide reasons to convince a sceptic that the circle is tangent to all three sides of the triangle.

Comment

There is a recording of the instructions on the CD-ROM (Imaginings) in case of need. You can use Figure 7.3b to check your visualisation.

For Euclid, it was necessary to justify this result on the basis of already established properties; starting from the intersection of two of the angle bisectors, he dropped perpendiculars to the sides of the triangle and used the conditions for congruence of various triangles. What else would you need to do to convince a sceptic?

Task 7.3.2 Particular In-circles

What is the radius of the in-circle of a '3, 4, 5' right-angled triangle?

What about a '5, 12, 13' triangle; a '7, 24, 25' triangle; an '8, 15, 17' triangle; a '6, 8, 10' triangle? Make a general conjecture about what you notice.

Comment

If you need to, draw a reasonably accurate diagram on squared paper.

The remainder of this sub-section provides the outlines for three different proofs of a formula for the radius of the in-circle of a right-angled triangle. One issue for consideration is what facility each proof requires in the reader. Notice what you have to do in each case to discern what is relevant, recognise relationships and perceive these as properties so as to appreciate the generality.

First Proof

The first proof is algebraic in nature. Start by assuming the in-circle has already been drawn inside a right-angled triangle. Denote that which is wanted but not yet known, namely the radius, by a letter, say r. The triangle can be split up into three triangles, each with a base on a side of the original triangle, and height r, as shown in Figure 7.3b.

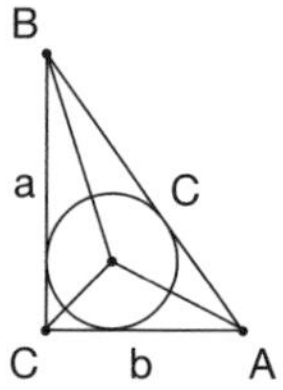

Figure 7.3b

Now use what is known about the radius: the area of each of the smaller triangles is half of one side of the triangle times the unknown radius r.

Denoting the sides of the right-triangle as a, b and c respectively with c as the hypotenuse, this yields

$r (a + b + c)/2 = ab/2$

from which r can be found algebraically.

Notice that if a, b, and c are integers, r will be rational. For the particular case of a 3, 4, 5 triangle, the radius turns out to be 1. For a 5, 12, 13, triangle, the radius turns out to be 2.

Second Proof

The second proof uses a different property, generalising a relationship that can be discerned in the diagram when you focus on tangents. Note from Figure 7.3b that the angle made at the right-angle is 45°, therefore r is equal to the length of the tangent from the right-angled vertex to the circle; then work out the lengths of the tangents to the circle from the other two vertices using the fact that tangents from a point are equal in length. Finally, use the fact that the hypotenuse is c and a different formula for r results, namely $r = (a + b + c)/2$. The two formulae are equivalent because $a^2 + b^2 = c^2$ as the triangle is right-angled.

Third Proof

The third proof is more of a geometrical demonstration. First the special case of 3, 4, 5 is tackled. Draw a circle, radius 1, centred on the right angle as in Figure 7.3c, part (i). Then draw a circle radius 2 centred on the vertex at the other end of the shortest side, as in part (ii) of Figure 7.3c. Finally, draw a circle radius 3 centred on the third vertex as in Figure 7.3c, part (iii).

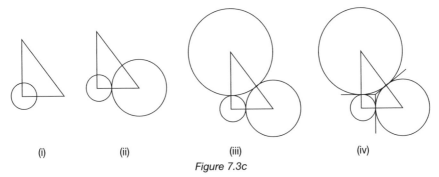

(i)   (ii)   (iii)   (iv)

Figure 7.3c

Surprisingly, the circles are mutually tangential. Draw in the common tangents as in Figure 7.3c, part (iv). These tangents are the radii of the circle you require and must all have the same length (being common tangents). Their length must be the same as the smallest radius because of the square that is formed due to the presence of the right-angle, so each has length 1.

This construction belongs in the category of being a special trick. It has a beauty and an efficiency but many people are left asking 'Where did that come from?', 'Would it work for other triangles?' and 'How were the radii 1, 2 and 3 decided upon?' In fact, the reasoning is based on discovering that you can predict the three relevant radii – they are the circles whose radii are the common tangents from each vertex in turn in the second proof.

Task 7.3.3 General In-circles

Draw a general triangle with its in-circle. Now discern some common tangents and, using these as radii, draw three mutually tangential circles each centred at one of the vertices of the triangle. What are the three radii in terms of the lengths of the sides of the triangle?

Comment

Use the '3, 4, 5' triangle to check your formulae.

The construction generalises, with radii $(a + b - c)/2$, $(a - b + c)/2$ and $(-a + b + c)/2$. Open and explore the interactive file '7c **In-centres**' to explore this generalisation. It is useful to denote by s the semi-perimeter of a triangle, that is, half the perimeter. Then the three radii can be expressed neatly as $s - a$, $s - b$ and $s - c$.

The third proof is undoubtedly mysterious: you have to know a great deal in order to 'pull it out of the bag'. But if learners are not aware of properties as independent facts to be used for reasoning, they are likely to find any proposed proof just as mysterious. In order to appreciate reasoning, it is essential to have recognised relationships that may exist between features in a diagram, and to have perceived these as properties that hold in general and not just in the particular diagram. Then and only then is it possible to reason on the basis of properties alone.

Some related problems can be found on the website NRICH(webref).

Ex-circles

The next task offers an opportunity to re-experience some of the same sort of reasoning in the context of ex-circles. The imagining Task 7.3.1 can be extended by allowing the circles to go outside the triangle as well. An *ex-circle* is a circle that is tangent to each of the sides of a triangle but whose centre lies outside the triangle itself.

Task 7.3.4 Ex-circles Challenge

Open the interactive file '7d **Ex-centres**'. Move the sliders and explain to yourself what you see.

Figure 7.3d shows part of the final image. Using a similar method to that used in Task 7.3.3, find the radii of the three circles in terms of *a, b*, and *c*, the sides of the triangle.

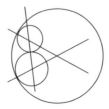

Figure 7.3d

On a copy of the final image from the interactive file, add the radii of the ex-circle to the two outer-most tangent points.

Looking at areas in two different ways, and using the ex-radius as an altitude where appropriate, find the radius of the ex-circle in terms of *a, b, c* and the area of the original triangle.

Comment

The radii of the circles centred at vertices will be $\dfrac{a+b+c}{2}$, $\dfrac{a+b-c}{2}$, $\dfrac{a-b+c}{2}$. The radii of the ex-circles will be of the form $\dfrac{area}{s-a}$

Note that the ex-circle and the in-circle on opposite sides of one side of the triangle are not tangent at the same points.

Were you aware of invariance and change? For example, no matter what shape the triangle, there are circles with centres at vertices, three ex-circles and one in-circle, all with interrelated radii?

Heron's Formula

The following description is intentionally brief to allow you to work at filling in the details. An imagining is available on the CD-ROM and there is also an interactive file '7e Heron'.

Think back to the first proof of the length of the radius of the in-circle. In this proof, the relationship $rs = area$ was derived. In the case of the right-angled triangle $rs = ab/2$.

The first part holds in any triangle. Denoting the area of the triangle by the Greek letter delta Δ (equivalent to letter D in English), gives the succinct relationship

$$\Delta = rs = r\,(a + b + c)/2.$$

The ex-radii have similar formulae. Denote the ex-radii by r_A, r_B and r_C, then

$$\Delta = rs = r_A\,(s - a) = r_B\,(s - b) = r_C\,(s - c).$$

Focus on the quadrilateral with vertices B, C, the in-centre I and the ex-centre E opposite angle A in Figure 7.3e. There is a right-angle between the pair of angle bisectors at B and also between the pair of angle bisectors at C. Consequently the opposite angles in the quadrilateral add up to 180°, which means the quadrilateral is cyclic: its vertices lie on a circle.

Dropping a perpendicular to side *a* from both the in-centre and the ex-centre, and noticing that these are symmetrically placed along side *a*, makes it possible to deduce that $(s - b)(s - c) = r\,r_A$ using the chord-product theorem.

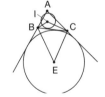

Figure 7.3e

It follows that if triangles have integer sides, they have rational in-radius and ex-radii. Integer-sided triangles can be constructed by joining together other integer sided triangles along a common edge. Integer-sided right-angled triangles form a starting point for such constructions. You might wish to explore whether all integer sided triangle can be constructed in this way, and whether there is any other integer-sided triangle besides the '3, 4, 5' triangle whose in-radius is 1.

Reflection 7.3

Reconstruct the path of development from the beginning of this section, starting with construction of the in-circle of a right-angled triangle, up to the presence of ex-circles and their relationships. What other pedagogic choices could have been made?

7.4 REASONING USING PYTHAGORAS' THEOREM

One of the reasons that Pythagoras' theorem is so important is that it provides a connection between measuring distances in a Cartesian co-ordinate grid (square roots of sums of squares of distances parallel to the two axes) and calculating areas. This section starts by developing a way to generate all Pythagorean triples (three integers that are the lengths of sides of a right-angled triangle), and hence all the points on a circle with rational radius, which themselves have rational co-ordinates. The second sub-section outlines Euclid's programme for calculating with areas leading to the class of all numbers that can be constructed using ruler and compass and starting with a unit length.

Pythagorean Triples

There is an ancient Babylonian tablet that has been interpreted as listing a large number of triples such as '3, 4, 5', '5, 12, 13', '7, 24, 25', in increasing order of the size of the smallest angle of the triangle. How might these be generated, and can you be sure you have found all possible integer-sided right-triangles?

Three formulae that have been used to generate Pythagorean triples are given in the tables that follow. These show methods of generation of triples – the proof that all cases are included is algebraic rather than geometric. The cases where the sides are integers are marked with $\star$.

Pythagorean Formula

This formula generates whole number Pythagorean triples for odd n.

n	$(n^2 - 1)/2$	$(n^2 + 1)/2$
1	0	1
2	1.5	2.5
3*	4	5
4	7.5	8.5
5*	12	13
6	17.5	18.5
7*	24	25

Platonic Formula

This formula generates whole number Pythagorean triples for even n.

n	$n^2/4 - 1$	$n^2/4 + 1$
1	⁻0.75	1.25
2	0	2
3	1.25	3.25
4*	3	5
5	5.25	7.25
6*	8	10
7	11.25	13.25
8*	15	17

Euclidean Formula

This formula generates a Pythagorean triple if $x > y$.

x	y	$x^2 - y^2$	$2xy$	$x^2 + y^2$
1	1	0	2	2
2*	1	3	4	5
3*	1	8	6	10
4*	1	15	8	17
5*	1	24	10	26
6*	1	35	12	37
2	2	0	8	8
3*	2	5	12	13
4*	2	12	16	20
5*	2	21	20	29
4*	3	7	24	25
5*	3	16	30	34
6*	3	27	36	45
7*	3	40	42	58

Task 7.4.1 Pythagorean Triples

Check, by squaring and adding, that each of the above three formulae does what it claims.

Comment

The point with co-ordinates

$$\left(\frac{2pq}{p^2 + q^2}, \frac{p^2 - q^2}{p^2 + q^2} \right)$$

lies on the circle with radius 1 centred at the origin. Furthermore, whenever p and q are integers (positive or negative now), this point has rational co-ordinates. Any point on the unit circle with rational co-ordinates must be of this form.

Pythagorean Construction

Now consider a more geometric problem. Given a rectangle that has integer sides, is it possible to construct another rectangle with double the area *and* double the perimeter that has rational sides?

Your first thought might be that doubling area requires scaling by a factor $\sqrt{2}$, but this does not give you rational side lengths. You have not been asked to preserve the proportions of the rectangle. If you focus on just one constraint, doubling the area is not difficult: you just make two copies of the original and join them on a common edge. However, this does not double the perimeter, so some modifications are required.

Starting with particular rectangles such as 3 by 4 rectangle (so its diagonal is also an integer), working with numbers arithmetically, and even using a spreadsheet to make computations faster does not shed much light. You might come across a 2 by 12 rectangle that has perimeter 28 and area 24 that double the perimeter and area of the 3 by 4. Starting with a 5 by 12 you might stumble over a 4 by 30; starting with an 8 by 15, you might discover a 6 by 40. Even so, it is not clear what the underlying structural relationships are that allow this to happen.

Thinking geometrically in terms of Pythagoras' theorem leads to a solution.

Start with two copies of a rectangle as in Figure 7.4a part (i).

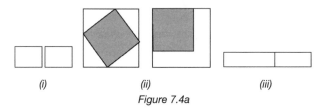

(i)   (ii)   (iii)

Figure 7.4a

Breaking them each into two right-angled triangles gives four triangles to arrange around a square hole (shaded) as in part (ii) of Figure 7.4a. The area of the large square is twice the area of the initial rectangle added to the area of the square hole. Rotate the hole inside the large square, to get the right-hand figure that leaves a *gnomon* shape of the correct area (double the original rectangle). This can be formed into a rectangle that has the correct double area (because area has been preserved in the operations performed) and, miraculously, the correct perimeter (double the original). In order to find such a construction, almost certainly someone noticed the relationship (doubled perimeter and area) and then posed the problem!

Task 7.4.2 Reviewing the Situation

Go through the proof above twice, first using the particular dimensions of the starting rectangle as 3 by 4, and then using general dimensions *a* by *b* based on a Pythagorean right-triangle *a, b, c*. Satisfy yourself that if the starting rectangle has integer sides and integer diagonal, the final rectangle will also have integer sides.

Comment

The reasoning works even if the diagonal is not an integer length, but the final rectangle will not then have integer length sides. It is worth going through the reasoning to see why the fact that the diagonal is an integer forces the final rectangle to be an integer. Following the reasoning through with a starting rectangle of 1 by 2, which is not Pythagorean, and calculating the area of the final rectangle, reveals the fact that $(3 - \sqrt{5})(3 + \sqrt{5}) = 4$.

Did you think to wonder if there is a similar construction for trebling the area and the perimeter?

One of the lesser known but important theorems of Euclid provides a construction of a square that has the same area as a given rectangle.

Task 7.4.3 Mean Geometry

Interpret Figure 7.4b as a sequence of moves to construct a square of equal area to a given rectangle. Provide reasons to justify each step.

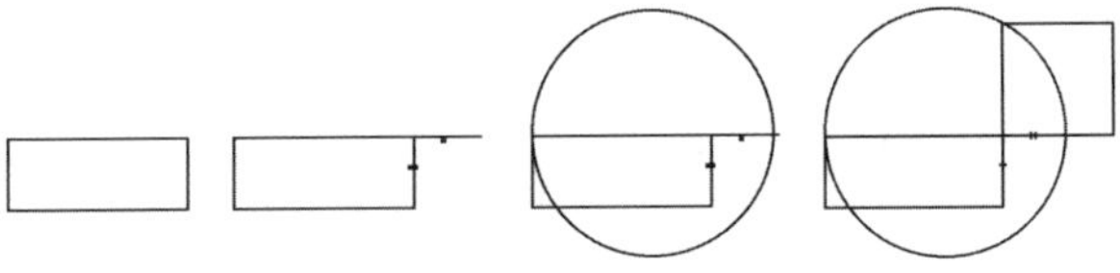

Figure 7.4b

Comment
Note that the length of the side of the square, being the square root of the product of the lengths of the sides of the original rectangle, is also their geometric mean.

Denoting the sides of the original rectangle by a and b, the circle has radius $(a + b)/2$ (the arithmetic mean). Using similar triangles and Pythagoras' theorem shows the vertical side of the square to have the desired length.

Euclid's geometry is the first known attempt to put a collection of mathematical results on a firm footing by being clear about what axioms or properties were assumed, and what properties were then proved. Another way of thinking about Euclid's geometry is as performing arithmetic geometrically.

Starting with a single *unit* line segment, segments with lengths that are whole number multiples of the unit can be formed. Using just these operations is sufficient to construct a collection of line segments corresponding to the positive whole numbers, also known as the counting numbers.

Starting with a unit segment and successively placing a unit segment at right angles to the previous one produces a Pythagorean spiral like the one in Figure 7.4c, with the radial lengths the square root of the positive integers. Adjusting the construction produces other interesting spirals (for example, placing the new unit segment at different ends of the preceding according to some pattern).

Given two line segments, it is not difficult to imagine how one might construct line segments representing the sum and the difference of the first two. Ruler and compass, or translation and (possibly rotation), is enough to do this.

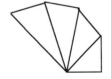

Figure 7.4c

Multiplication and division of line segments is not immediately so obvious, but they are made possible using Thales' theorem on ratios (see Chapter 2). It is not clear whether Euclid even conceived of the mathematical task of adding ratios, even though he could add line segments.

Constructing square roots is straightforward using the construction in Task 7.4.3. This means that it is possible to construct segments corresponding to such numbers as

$$1 + 2\sqrt{3},\, 1 + 2\sqrt{3 - \sqrt{4/5}} \text{ and } \frac{1 + 2\sqrt{3 - \sqrt{4/5}}}{1 + 2\sqrt{3}}.$$

Indeed, any number formed by using addition, multiplication, division and square roots (of positive quantities) can be constructed as a segment. This turns out to be the set of numbers that can be constructed using just an unmarked straight edge and a pair of compasses.

Considering areas, even addition is not at all clear. Euclid did know that any polygonal region can be decomposed into triangles. He also proved theorems that show how a triangle can be converted to a rectangle of the same area, and then how a rectangle can be converted to a square of the same area. Pythagoras's theorem comes to the rescue, because it shows how to add areas of squares. The final square can then be recomposed as a triangle or some other shape by reversing the operations that led to the two squares in the first place. Euclid did not include anything about multiplying areas, presumably because there is no physical object whose measure requires four dimensions.

Geometry offers an entry to ways of constructing chains of reasoning and tying these arguments back to real pictures in the mind. One way of working with geometry is to start with the geometry and stay with it, but there is a second aspect; geometry is a way into number. In no way should we underestimate the power of being able to think with an unstructured number line. Nor should we underestimate the way a visual starting point can captivate and intrigue. Diophantus, a mathematician who lived and worked in Alexandria about 300 years after Euclid, produced a number of books, some of which survived the upheavals of the next thousand years. Diophantus posed several 'whole number' problems. Copies of his books were still circulating amongst mathematicians a thousand years later. The famous remark by Fermat that he had proved that $A^n + B^n = C^n$ only has whole number A, B and C solutions when $n = 2$ was written in the margin of his copy of Diophantus alongside 'this margin is too small to hold the proof'. This problem stimulated amateur mathematicians for the next 200 years and was only solved by Andrew Wiles at the end of the twentieth century. Perhaps one lesson to take from this story is that whole number starting points can sometimes make hard problems accessible.

Reflection 7.4

Review mentally the different roles that Pythagoras' theorem plays in mathematics.

Review the features that different proofs of the same theorem might offer, and how these might affect a choice of what to offer learners.

7.5 PEDAGOGIC PERSPECTIVES

The pedagogic approach in this chapter has been to stick with just a few results, but to use them to explore the nature of reasoning and proof. This is more in line with the approach to teaching geometry in Japan and Hungary than it is with that commonly found in England. The curriculum specifies a variety of results, or 'geometrical facts', whereas what is most important is that learners realise the fact that there are geometrical facts, rather than internalising specific facts, learning to reason, through shifting attention from specific relationships to properties, and then to reasoning solely on the basis of agreed or accepted properties. Throughout what matters most is that learners' powers are engaged and developed, and that they encounter persistent mathematical themes such as invariance within change. The pedagogic task is to take a number of important mathematical processes and themes, and then select those items of content to work with in depth that will allow pupils to get a feel that they have ownership of working and thinking visually.

Perhaps the most important power of all is the power to imagine, and to express what is imagined in gestures, movements, diagrams, words and symbols. You have frequently been invited to imagine. Certainly both before and after using dynamic geometry software it is important to imagine what is possible, and to re-imagine what happened. When working with others it is valuable to try to imagine what they are describing and to learn to describe accurately what you are imagining. This is how the power to imagine is developed.

Asked to imagine a square, attention may be attracted to many different features: vertices, sides, angles, parallelism, equality of length, equality of angles, but also colour, thickness, orientation and taste. Some aspects are useful mathematically, others are not; some aspects are to be discerned, others are relationships to recognise; others are properties that are being exemplified. The strategy of getting learners to 'say what you see' and to negotiate these different features is an effective way of directing attention to mathematically significant features and aspects, and of directing attention to recognising relationships and then perceiving properties.

Many pupils find geometry, in the early stages, more accessible than algebra. This accessibility is tied in to being more comfortable in handling materials, reading and working with pictures and figures than manipulating abstract symbols. But pedagogically a teacher needs to move pupils on from enjoying and being comfortable to becoming involved and effective as a geometer. A key to becoming sharper in working with images is contained in the phrase: say what you see – watch what you do.

The gap between working on your own with course materials and working as a teacher with pupils is very sharp here. A pedagogic strategy with a group of pupils, is to ask individuals to 'say what they see'. There are further strategies, like insisting that what is seen is described in words and not by approaching the white board and waving one's arms about. The teacher's job in setting up this situation is to provide the scaffolding for discussing what is seen, but also to fade this scaffolding as the discussion progresses. However, the teacher needs to be watchful to ensure that the removal of the scaffolding does not result in a deterioration of the quality of the geometric seeing and reasoning.

A second attribute of proof in geometry is concerned with developing chains of reasoning. Learners need to experience reasoning (the giving of reasons for why assertions are true), and to try to provide reasons for themselves, convincing themselves, then a friend, then a sceptic. This was why sometimes you were given proofs to follow and sometimes invited to reason for yourself.

This chapter sits within measurement because every thing discussed either involves some sort of measurement (as in lengths, angles, and radii) or plays an important role in measurement. Most of the work required you to reason and not measure. This is central to classical Euclidean geometry. Aristarchos, in 250 BCE, used Euclidean geometry to measure the diameters of the sun and the moon and to estimate the relative distances of the two from the earth. His observations and results led him to postulate that the sun was the centre of our universe and not the earth. This story is a good example of the interplay between geometrical deduction and producing a measurement in a case where it is not possible to lay down a ruler and read off a result.

Reflection 7.5

Next time you see some mathematics that you find off-putting, try approaching it using 'say what you see' and notice the effect on you.

8 Visualising

This chapter develops links between the square of a number, the area of a square, the path traced by a point that moves equidistant from a fixed point and a fixed line, a quadratic expression such as $x^2 - 1$ or $(x + 2)(x - 1)$ and the graph of a quadratic expression.

The unifying theme of this chapter is how geometrical thinking can be developed to assist in the visualisation of mathematical ideas and, in particular, developing the concept of locus to gain an understanding of graphs.

An important step was made by René Descartes developing the notion of co-ordinates in the seventeenth century. Before Descartes, mathematicians thought of curves as defined by intrinsic properties: a circle is the set of point a fixed distance from a fixed point, a parabola is the set of points equal distance from a fixed point and a fixed line, and so on. Descartes wrote about a fevered dream in which he realised that instead of studying curves by their intrinsic properties, it would be possible to use the power of algebra by specifying curves as an equation relative to a co-ordinate system. This insight transformed the way curves were studied.

This chapter makes links between what we have come to think of as distinct branches of mathematics and shows how these branches of mathematics have applications in other areas of study.

8.1 DYNAMIC GRAPHS

When you hear or see the word *graph*, what comes to mind? Most likely what the word conjures up is a static picture of co-ordinate axes and a curve. But the static is merely a snapshot from a dynamic 'film'. This is brought out in the idea of a *locus*.

Locus

The term *locus* comes from the Latin for place, and means the set of points that satisfy some condition.

Task 8.1.1 First Locus

Imagine two straight lines. Now imagine a point that moves so that it is always the same shortest distance from each straight line. Where can it get to (that is, what is its locus)?

Now make the two initial straight lines into perpendicular axes. Focus attention on some point equidistant from the two axes. What relationship must there be between the co-ordinates of the point?

Now imagine a point moving along the *x*-axis. Imagine that as it moves, another point accompanies it. This point has its *y*-co-ordinate the same as its *x*-co-ordinate. What path does it follow?

Comment

Did you notice that the point can be on either of two straight lines unless the original lines are parallel? What is the relationship between those two lines, and what is their relationship with the original straight lines?

The co-ordinates are either equal, or one is the negative of the other. Forcing the co-ordinates to be equal can be seen as a static condition leading to a locus, and also as a dynamic condition that restricts the freedom of a point that moves as the *x*-co-ordinate changes.

The interactive file '**8a As far as**' allows you to alter two straight lines and a ratio. The file displays the locus of points whose distances from the two straight lines are in the chosen ratio (see Figure 8.1a). To convince yourself, or someone else, that the locus is correctly drawn, you will need to work out how the dashed lines are related to the ratio, and then to draw in a general point on the locus, drop perpendiculars, and use similar triangles.

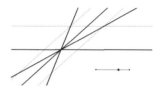

Figure 8.1a

Task 8.1.2 Dynamic First Loci

Figure 8.1b suggests a way of making a dynamic geometry software file (see '**8b Straight Lines**') to display the locus of points whose co-ordinates satisfy the constraint that $y = mx + b$ where m and b are governed by their positions on the *y*-axis. Explain why the co-ordinates of P are indeed $(x, mx + b)$.

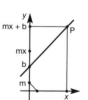

Figure 8.1b

Comment

The best way to get a sense of what is going on is to drag one of the objects to see what changes and what stays the same. Most importantly, dragging the *x* point reveals the locus of the point P as *x* changes. Dragging the *b* point up and down translates the locus; dragging the *m* point changes the slope of the graph.

Thales' theorem explains why the end of the line parallel to the segment from *m* to the unit point $(1, 0)$ is *mx* from the origin. The segment marked *b* is a translation of the point *b* from the origin in order to add on *b*.

Circle Graphs

Traditionally, graphs have been derived from tables of values, with the intention that the symbolic expression gains meaning by being associated with the graph. Spreadsheets reinforce this by moving from a table to a graph as a final static product. Using dynamic geometry software, graphs can be seen both as the depiction of a movement and an opportunity to get a sense of the role played by slope and intercept through seeing the effect of changes in these.

Task 8.1.3 Circle Graphs

Figure 8.1c shows three graphs. One is the graph of area of a circle against the perimeter, one is the area against the radius and one is the perimeter against the radius. Which is which, and why?

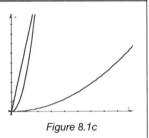

Figure 8.1c

Comment
The interactive file '**8c Circle Graphs**' provides a dynamic environment for generating these graphs. Getting learners to conjecture the shape of graphs reinforces their awareness of relationships and helps them become aware of these relationships as properties. Some suggested graphs are:

- perimeters of shapes (triangles, parallelograms, trapezia) plotted against an edge length by which the shape is scaled
- perimeters of shapes plotted against one edge that changes while the rest remain the same
- areas of shapes as an edge changes.

Reflection 8.1

How does thinking of a graph as dynamically generated by a moving point alter or enrich your sense of what a graph is?

8.2 PROJECTIONS

In Chapter 4 you considered projections of a solid as used in technical drawings. Projections are more widely used than might at first appear. The x and y co-ordinates of a point P are the projections of P on the respective axes, so map reference systems are using projections to co-ordinate points on the map. Co-ordinates 'co-ordinate' the location of a point by reference to two or more independent axes, so that, for example, the movement of a point can be tracked by recording its projections on two different axes. Navigators and map-makers have faced this problem ever since people went to sea and travelled long distances. Before connecting trigonometric ratios to projections in the next section, this section considers briefly the issue of map-making as a projection problem.

Figures 8.2a and 8.2b show two different representations of the mouth of the River Tyne. It is possible to identify on the map (Figure 8.2a) the location of the camera that took the photograph (Figure 8.2b).

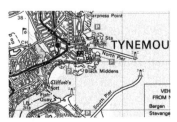

Figure 8.2a

Figure 8.2b

Such representations are familiar, though it is not often that they need to be co-ordinated. In the case of the relatively small geographical area, shown here in different ways, the map gives a clear picture of what it is like on the ground: relative distances closely match distances measured on site.

Representing the Earth

A problem arises when you want to map continent-sized tracts of land. The curvature of the Earth affects the apparent distance in relation to the measured distance, because the Earth is approximately spherical: in fact, it is an *oblate* sphere, being slightly squashed at the Poles.

Imagine peeling an orange and attempting to flatten a piece of peel onto a plane surface. Unless you take a very small piece you will have to make a tear in it. If you do take a very small piece it will indeed be easy to think of it as flat – just in the way that small portions of the Earth's surface such as the mouth of the River Tyne can be represented by flat surfaces (maps). How, though, can the complete surface of a sphere be represented on a plane surface?

In Chapter 5, spherical co-ordinates were introduced in terms of latitude and longitude. Figure 8.2c (i) shows a cross section through the centre of the Earth and a point P on the surface of the Earth, north of the Equator. Part (ii) shows a projection onto a plane through the North and South Poles; part (iii) shows a projection onto the equatorial plane.

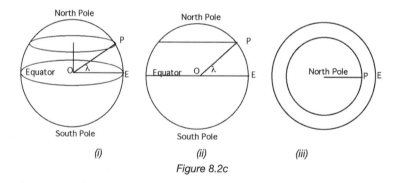

Figure 8.2c

Task 8.2.1 Changing Latitude

Sketch a graph of how the ratio of the circumference *C* of the circle of latitude to the equatorial circumference changes with latitude.

Comment
Did you think to sketch the circumference of a circle of latitude against the latitude before going for the ratio? How are the two related? There is a dynamic geometry software file '**8d Latitude and Longitude**' to help you think about latitude and longitude.

Since the radius of the circle of latitude at latitude λ is $r \cos \lambda$ (λ is read 'lambda') the ratio of the circumferences is $\cos \lambda$, which might have surprised you. Many people expect a straight line. Think about extreme cases: at latitude 0° the circle of latitude is the equatorial plane so the ratio is 1. At the North Pole the latitude is 90° and the radius, and therefore the ratio, is zero.

Ideally, in trying to map the surface of a sphere onto a flat plane of paper, you would want meridians (half circles of longitude) and parallels of latitude to preserve (keep invariant) the following relationships:

- parallels of latitude are parallel
- parallels are equally spaced along meridians
- meridians are equally spaced along parallels
- meridians and parallels intersect at right angles
- quadrilaterals formed by the same two parallels and same two meridians have the same areas
- area scale is uniform
- distance scale is uniform.

The term 'map projection' is used when this spherical grid is transformed onto a rectangular grid. This allows geographic co-ordinates (latitude and longitude) to be transformed into Cartesian (x, y) co-ordinates. Some of these can be geometrically constructed: you can think about obtaining a map of a spherical surface on a plane surface by imagining a light source somewhere in relation to a transparent globe and projecting shadows of the meridians, parallels and the geographic features of the Earth onto a sheet of paper. This sheet of paper may be wrapped round the Earth tangential to the globe at the Equator. As Figure 8.2d shows, the resulting map will be very different depending on where the light source is placed.

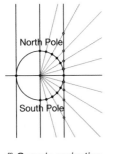

(i) Gnomic projection

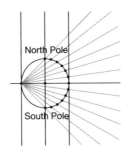

(ii) Stereographic projection

Figure 8.2d

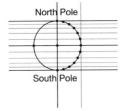

(iii) Orthographic projection

Gnomic Projection

A *gnomic* projection uses a cylinder wrapped around the sphere (Figure 8.2e overleaf). Think of a light source at the centre of the sphere.

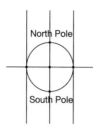

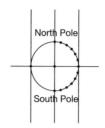

(i) This view shows a cross section through the centre of the Earth. The vertical lines tangential to the circle are the cut edges of a cylinder wrapped around the Earth.

(ii) Consider points on parallels of latitude every 15°.

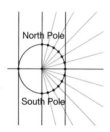

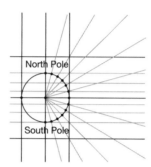

(iii) Project rays from the centre of the Earth through the points on the Earth's surface

(iv) This gives parallel grid lines every 15° of latitude.

Figure 8.2e

Task 8.2.2a Gnomic Projection

What remains invariant under this projection? What does not?

Comment
The order of points is preserved going round each circle of latitude, but the relative distances are changed, because the farther you are from the reference point the larger the angle and so the greater the distortion. Each ray in the diagram cuts the Earth so that the arc lengths are the same – subtending an angle of 15° at the centre of the Earth. These equal distances are represented by parallel lines of the grid shown.

Consider a point P that can move around the circumference of a meridian. Think about it starting at the Equator and consider its angle of latitude $\angle EOP(\lambda)$ increasing from 0° to 90° (when it reaches the North Pole) – see Figure 8.2f.

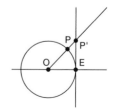

Figure 8.2f

Task 8.2.2b Latitude Projection

Sketch a graph of how the distance of P'E varies as P moves round the earth. Consider how this graph develops as P moves from the Equator to the North Pole, (as the latitude varies from 0° to 90°).

The mapped or projected height of P above the equator E is $r \tan \lambda$. The graph of $\tan \lambda$ will be generated dynamically in the next section. Knowledge of how trigonometric functions vary is clearly important for considering how to map the Earth's surface effectively.

Normal Azimuthal Projection

One type of stereographic map projection is the Normal Azimuthal projection. In this the projection is made to a plane that is tangential to the sphere at a single point – in normal (or polar) cases the point is either the North or South Pole, as shown in Figure 8.2g.

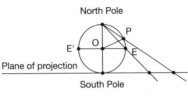

Figure 8.2g

Task 8.2.3 Azimuthal Projection

What happens as P moves between E and N (North Pole). What stays the same and what changes as far as mapping is concerned? The interactive file '**8e Stereographic Projection**' may help.

Comment
In projections of this type meridians appear as radial straight lines and the parallels (of latitude) as concentric circles. The spacing of the parallels increases rapidly with distance from the pole that is in the plane of projection. It turns out that angles are preserved even though distances change.

This is the oldest type of map projection and is accredited to Thales. Because great circles are straight lines on the map, it has been used to aid navigation.

Central and Polar Projections

Task 8.2.4 Central Projection

Project point P on the Earth's surface by a ray through P from the centre of the Earth, onto a plane tangent to the Earth at the North Pole. Express the distance of the projected point on the 'map' from the North Pole in terms of the angle of latitude of P. How does this projected length vary as P moves from the Equator to the North Pole?

Comment
Draw your own diagram (most useful might be an elevation of the Earth showing the tangent plane as a straight line). The distance from the projected point to the North Pole is $r \cot \lambda$. Care is needed to distinguish between measuring angles in degrees and in radians. Degrees are fine until you want to measure arc lengths, when it is most convenient to use radians. If λ is the latitude in degrees, then $2\pi \lambda/360°$ gives the latitude in radians.

An alternative is to central project from the Pole opposite to the tangent plane, so if the North Pole is the tangent point and at the centre of the map, the South Pole is the point of projection.

Task 8.2.5 Polar Projection

Project point P on the Earth's surface by a ray through P from the South Pole, onto a plane tangent to the Earth at the North Pole. Express the distance of the projected point on the 'map' from the North Pole in terms of the angle of latitude of P. How does this projected length vary as P moves from the Equator to the North Pole?

Comment

This time you need to use the fact that the angle subtended by a chord of a circle at the centre of the circle is double the angle subtended at the circumference. The result is that the distance is now $r \tan (\pi/4 - \lambda/2) = r \cot (\pi/4 + \lambda/2)$.

Mercator Projection

The last projection to be considered is perhaps the most famous: the Mercator projection. In fact it is not a true geometrical projection at all; it preserves angles (directions) but distances are distorted.

Mercator, was born in Flanders, which is now part of Belgium, in 1512. He left no written record of his method except the following brief explanation printed on his map:

> In making this representation of the world, we had to spread on a plane the surface of the sphere in such a way that the positions of places shall correspond on all sides with each other both in true direction and in distance … With this intention we had to employ a new proportion and a new arrangement of the meridians with reference to the parallels … For these reasons we have progressively increased the degrees of latitude towards each pole in proportion to the lengthening of the parallels with reference to the equator.
>
> (Maor, 1998, p. 173)

What Mercator did was to divide the globe into a large number of meridians. Each section between two adjacent meridians was than expanded so as to fill a rectangle of the same width as the equatorial width. The effect is to stretch out the lines of latitude so that directions on the map (and hence angles) are (more or less) preserved. Mercator realised that conserving angle meant conserving length was not possible, and abandoned attempts to preserve both.

Although Mercator did not explain the underlying mathematics, by 1599 the English mathematician Edward Wright wrote that 'by perpetuall addition of the Secantes answerable to the latitude of each parallel unto the summe compounded of all former secantes [it is possible to] make a table which shall truly shew the points of latitude in the Meridians'. (Maor, 1998, p. 174)

To see what he meant, consider the lozenge shape between two adjacent meridians, which themselves are $R\theta$ apart at the Equator (see Figure 8.2h). At a height measured by the latitude λ the width of the lozenge on the Earth is, as Task 8.2.2b revealed,

$R\theta \cos \lambda$. Since this needs to be scaled up to $R\theta$ for use on the map, a scale factor of $1/\cos \lambda = \sec \lambda$ is required. In order to preserve directions, that is, angles or slopes, very close to that latitude, distances along the meridian have to be scaled similarly.

To work out the general effect requires the use of the calculus, which is exactly what calculus was developed to do, some 100 years after Edward Wright was making his maps. What Wright could do was to approximate by dividing the sphere up into a series of lines of latitude close together (1 minute of arc = 1/60th of a degree), and then add together the secants of the angles to work out the scaling required.

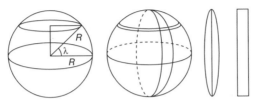

Figure 8.2h

Projections have proved useful to artists as well as to map makers. Bridget Riley is a British artist who often uses geometrical design in her work. For example, in her work *Movement in Squares* (Riley 1961), the illusion of a curved surface can be achieved by taking the net of a cylinder, drawing a regular grid of squares, forming the cylinder, then projecting that onto a tangent plane, as suggested in part (i) of Figure 8.2i. The full projection is shown in part (ii).

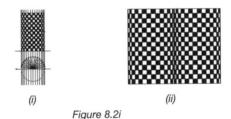

(i)   *(ii)*

Figure 8.2i

Reflection 8.2

It is often claimed that encountering ideas in a reasonably familiar context helps in appreciating the significance of those ideas. Having met *tan* and *cos* as projections involved in map-making, has this helped you? Can you mentally reconstruct the significance of *tan* and *sin* for map projections?

8.3 TRIGONOMETRIC GRAPHS

Looking again at the diagrams for the trigonometric ratios (see Figure 8.3a), and thinking in the language of projections, the cosine gives the projection onto the *x*-axis and the sine gives the projection onto the *y*-axis. In Figure 8.3a, the angle θ has been omitted from the trigonometric expressions so as to save space.

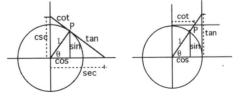

Figure 8.3a

Consequently the Cartesian co-ordinates of a point P in the plane can be given as $(r \cos \theta, r \sin \theta)$ where r is the distance of the point from the origin. The projection of P onto the tangent to the circle at the x-axis results in a point with co-ordinates $(1, \tan \theta)$, and the cotangent is the corresponding projection of P onto the tangent to the circle at the y-axis, so its co-ordinates are $(\cot \theta, 1)$.

Task 8.3.1 Graphing Sin and Tan

Look at Figure 8.3b, then imagine point P moving around the circle at a uniform speed.

Pay attention to the projection of P on the y-axis as P moves. Where does it change direction? Where does it seem to move fastest, slowest?

Now do the same for the projection of P from the origin onto the tangent at $(1, 0)$.

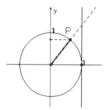

Figure 8.3b

Comment

If people act out the projections, it helps to establish both the image and a kinaesthetic sense of the changes in the lengths of the projections. One person walks around a circle, another walks up and down the y-axis following the orthogonal projection, and one person walks up and down the tangent following the projection from the origin. You can add another person to follow the projection of P on the x-axis as well.

To turn these changing lengths into graphs involves tracking the changes in lengths as the length of the arc between P and $(1, 0)$ changes. Dynamic geometry software makes it is possible to construct a dynamic version of these projections and projections for the other trigonometric functions. The file '**8f Trig Graphs**' provides access to these. The important thing is to get a sense of how the projections vary as P moves around the circle. Watching how the projection changes as P is dragged or animated round the circle *before* constructing the locus helps to establish the mental image, even if your conjecture is not quite right.

Task 8.3.2 Pythagoras' Theorem Revisited

By contemplating the right-angled triangles in Figure 8.3a, explain:

why $(\sin \theta)^2 + (\cos \theta)^2 = 1$,

why $(\tan \theta)^2 + 1 = (\sec \theta)^2$

and why $(\cot \theta)^2 + 1 = (\csc \theta)^2$

not just for the particular θ shown but for all possible θ.

Comment

Each is an expression of Pythagoras' theorem. Each can be obtained from the others by expressing the various trigonometric functions in terms of $\sin \theta$ and $\cos \theta$.

Relating Trigonometric Functions

Because of the relationships between the six trigonometric functions there are all sorts of further relationships, many of which can be introduced by setting them as plotting tasks. Surprise at the graphs can motivate a desire to know why the relationship always holds.

Task 8.3.3 Comparing Graphs

Graph the following sets of functions on a graphics calculator, spreadsheet or dynamic geometry software. Articulate each phenomenon as a conjectured relationship amongst the trigonometric functions.

$\sin(\pi/2 - x)$ and $\cos x$

$\sin^2 x$, $\cos^2 x$, and $\sin^2 x + \cos^2 x$

$\sin 2x / \cos x$ and $\sin x$

$1 + \cos 2x$ and $2\sin^2 x$

$\tan 2x$ and $2 \tan x / (1 + \tan^2 x)$

Comment
Did the change from θ to x make any difference to your thinking? With one exception, in each case you will find that the graphs appear to coincide. The first can be justified by looking at a right-angled triangle and noting the location of $\pi/2 - x$. The second is, of course, Pythagoras' theorem again. The third and fourth can be justified by using a more general result from the next task. The last can be justified by using the third and fourth and the definition of tan in terms of sin and cos. Exploring graphs more widely, such as $x \sin x$, $x^2 \sin x$, $\sin x^2$ and even $x^2 \sin (1/x)$ helps develop a collection of familiar functions.

As a final relationship that uses co-ordinates rather than dynamic images, the next task shows how to connect the sine and cosine of sums and differences of pairs of angles in terms of the original angles.

Task 8.3.4 Angle Sum

Examine Figure 8.3c. Two angles are labelled A and B. six points are labelled O, P, Q, R, S, and T. Locate right-angled triangles that have angles A, B and A + B respectively.

Use those triangles to express sin (A + B) as a ratio. Write the numerator, and hence the ratio, as a sum. The task now is to reinterpret those lengths using lengths relevant to the sin and cos of angles A and B.

Note that multiplying one of the ratios in the sum by OP/OP and the other by QP/QP makes it possible to use the sin and cos of A and B.

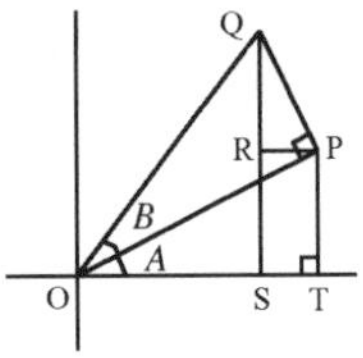

Figure 8.3c

Having found that $\sin (A + B) = \sin A \cos B + \cos A \sin B$, it is possible to do the same sort of calculations to find $\cos (A + B)$ and $\tan (A + B)$, but it is also possible to use the fact that $\cos \theta = \sin(\pi/2 - \theta)$ to deduce the result for cos, and then use both results to deduce the result for tan. Another approach that involves a little less algebra is used in Chapter 11.

Lissajous Figures

The final task is an opportunity to use co-ordinates and sin and cos both to explain and to produce a phenomenon.

In many of the activities of this chapter you have seen how geometrical understanding can be adapted to develop graphs in which one variable is plotted against another. In most cases you have explored how the variable plotted on the horizontal axis may be considered to increase through all values, for example when measuring how the area of a circle varies with its radius. However, graphs can be used to explore situations where each of two variables are changing differently with time.

Task 8.3.5 Lissajous Figures

Imagine a circle, and a point moving at uniform speed around that circle. Now imagine a second circle with a point moving around it also at uniform speed, but not necessarily the same as speed as the first point (nor even the same radius). Imagine a horizontal line through the first point as it moves, and a vertical line through the second point as it moves. Where these two lines cross, mark a point P. To ask what sort of paths P might follow as the points move is very ambitious. It is probably best to start with something simple: circles the same radius, points moving at the same speed and in the same direction. What variations will occur depending on how the two circles are placed, and how the two points are placed on their circles initially?

Comment
Such curves were first investigated in the mid-1800s by Jules Lissajous – since then they have often been known as Lissajous Figures. Originally they were drawn using a mechanical device called a harmonograph (Figure 8.3d) based on two circular pendulums. It is possible to develop these figures using an oscilloscope in which oscillatory waves are connected across both horizontal and vertical directions; however, it is more usual in an oscilloscope to have a time function in the horizontal direction.

Figure 8.3d

Task 8.3.6 Undoing a Lissajous

In how many different ways can the circles and their points and speeds be chosen so that the Lissajous Figure is a circle, or a figure of eight?

Comment
There is an interactive file '**8g Lissajous**' to support your work on this task.

There is a great deal of worthwhile work to be done developing a sense of the behaviour of the sine, cosine and tangent functions. Thinking in terms of the graphs generated dynamically provides a much more lasting impression than simply displaying the graphs as finished objects. Acting out the projections adds a body-based sense to the imagery and the intellectual knowledge.

Reflection 8.3

What did you find most effective in appreciating the behaviour and shape of the graphs of sine, cosine and tangent: acting out the projections, imagining the projections or using the interactive files? In what ways does the work on map projections support your appreciation of trigonometric functions?

8.4 MORE LOCI

In Chapter 4 you met some initial ideas about how considering the path of a point that is constrained to move in a certain way can give rise to paths that are geometrical figures. In this section you have the opportunity to consider some other well-known curves that can be defined as the path of a point that is constrained to move in a particular way and see how such curves can be used to model real structures and situations.

Circles

The most well-known example of this type is a point that moves so that it is always the same distance from a fixed point, giving a circle.

Task 8.4.1 Apollonius Circles

Mark two distinct points F and G. Consider the locus of points P that are twice as far from F as they are from G. Why is it a circle?

Comment
Apollonius was born in what is now Turkey around 262 BCE and is best known for his study of conic sections. The easiest way to see why the locus is a circle is to use co-ordinates, Pythagoras' theorem and some algebra (some squaring is involved). What if the ratio of 'twice' was changed to something else?

The interactive file '**8h Apollonius**' uses Thales' theorem (hidden from view) for its construction and will help you explore this task.

Parabolas

Another commonly occurring example of this type is a point that moves so that it is always the same distance from a fixed point as it is from a fixed line.

Task 8.4.2 Parabolic Locus

Imagine a straight line, and a point that is fixed somewhere away from the line. Now imagine a point moving so that it is always the same distance from the fixed point as it is from the fixed line.

What shape is the locus?

Comment
If you are not familiar with this construction then it will help to use dynamic geometry software. To construct the locus, use a straight line parallel to the bottom of the screen, and a

point somewhere above it. Any point on the locus must be the same distance from the line as from the point. Construct a point P on the line (this will generate the locus). Draw a line through P perpendicular to the fixed line. Then the locus point corresponding to the position of P on the fixed line is somewhere on this perpendicular. Indeed, it must be equidistant from P and from the fixed point F. Since the locus point must be on the perpendicular bisector of FP as well, where that cuts the perpendicular through P must be the locus point.

The parabola is a very special curve. It is used to make satellite dishes because of its ray-focusing property (see below), as well as bridges. Most of all, it is the path followed by a fountain of water, or any ball thrown or kicked into the air

There are a number of structures that have a parabolic shape with which you may be familiar – one of these is shown in Figure 8.4a. There are structural reasons why such shapes are used in building bridges – parabolic arches are good at minimising stresses and strains.

The photograph in Figure 8.4a has been imported into an interactive file '**8i Tyne Bridge**' so that you can try to construct a parabola that fits it.

Figure 8.4a

Task 8.4.3 Tyne Bridge

Where in the photograph are the fixed line (known as the *directrix*) and the fixed point (the *focus*) likely to be? By constructing a parabola as suggested above, and then moving the line and the focus, it is possible to get a pretty good match.

There are other ways of geometrically defining a parabola – as an envelope of straight lines. The next two tasks provide an opportunity to relate a paper-folding action to the defining property of the parabola.

Task 8.4.4a Folding a Parabola

Take a sheet of tracing paper and mark a straight line across it. Mark a point, F, somewhere on the paper away from the straight line. Now fold the paper so that a point on the straight line lies on top of F. Doing this many times produces a family of straight creases that envelope a parabola. Why does this happen?

Comment

To model the folding, using dynamic geometry software, you need to find the crease line when a point P on the fixed line is folded to lie on top of the fixed point. This is the perpendicular bisector of PF (convince yourself). The locus construction will then produce a family of such crease lines and reveal a parabola.

Task 8.4.4b Folding Another Parabola

As before, take a sheet of paper and mark a straight line across it. Mark a point F somewhere away from the line. Through any point P on the straight line, construct a line perpendicular to the line FP. Doing this many times produces a family of lines that envelope a parabola. Why does this happen?

Comment
What other aspects could you vary, and still get a parabola?

Task 8.4.5 Fountains

Open the file '**8j Fountain**' and fit a parabolic envelope to the jet of water issuing from the image of a fountain. One possibility is shown in Figure 8.4b taken in the garden of the Alhambra in Spain.

Comment
Being able to import images into dynamic geometry software means that you are able to explore the geometry of any digital photograph you may have access to, either through photographs or on the web.

Figure 8.4b

You will no doubt have noticed the parabolic shape of satellite television receivers. It is interesting to consider an important geometric property of the parabola that makes it so useful in a number of such situations in the real world. Rays that are travelling parallel to the axis of symmetry of a parabola are reflected to the focal point (Figure 8.4c(i)).

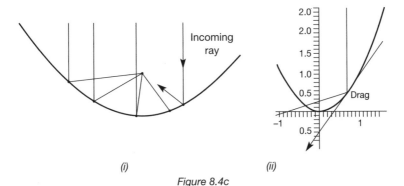

Figure 8.4c

This property is used in satellite dishes to focus the signal to the transponder. It is used in reverse in some car headlights: a light source at the focal point can be used with a parabolic reflector to produce a strong beam of light with parallel rays.

The interactive file '**8k Parabolic Reflector**' demonstrates this property. (See Figure 8.4c (ii)). This file allows you to drag a tangent along a parabola and see that a ray parallel to its axis will be reflected to the focus of the parabola.

Ellipses

Another curve that you can explore as the path of a point that is constrained to move in a certain way is the ellipse. You may have already met one method as it is commonly told to pupils at school. It is based on the fact that if you pin a piece of string at two points such that the length of string is longer than the distance between the two points and you move the piece of string so that it is always taut, the path of the extreme point is that of an ellipse.

Task 8.4.6 Elliptical Locus

Mark two points. Construct the locus of points for which the sum of the distances to the two fixed points is constant.

Draw a circle and mark a point inside the circle. Now draw a circle that is tangent to the fixed circle and passes through the fixed point. What is the locus of centres of these circles?

Comment

A dynamic geometry software makes exploration pleasurable as well as easier. Having thought deeply about how to make the construction, the 'fixed' points and circle can be varied to see the effect on the locus. Convincing yourself that the second locus satisfies the construction principle of the first locus requires locating a second suitable point (there are not many to choose from!) and drawing a particular case with various radii in order to 'see' the underlying structure.

An ellipse is also the locus of points for which the ratio of distances from a fixed point and a fixed line is a constant less than 1 (this ratio is known as the *eccentricity*). If you allow the ratio to be greater than 1, you get a different curve, a hyperbola. All three curves, the parabola, ellipse and hyperbola can be seen as sections of a cone cut by a plane. The parabola occupies a position between the family of ellipses (which includes circles as a special case) and the family of hyperbolas.

An ellipse can also be developed as an envelope of straight lines as the next task shows.

Task 8.4.7 An Elliptical Envelope

Draw a circle and choose a point F somewhere inside the circle away from the centre. For any point P on the circle, join P to F and construct the perpendicular to PF through P. Make a crease along that line. Why do the lines envelope an ellipse?

Cut a circle out of a piece of paper, or obtain a circular filter paper. Mark a point F away from the centre. Fold the paper so that the edge of the circle lies on top of F and make a crease. Why does the family of creases envelope an ellipse?

Comment

Having discerned a specific crease, locate the objects (two points perhaps) needed to reason about an ellipse. Then draw in other features that relate the crease to those features and use properties of folding to support your reasoning.

Figure 8.4d shows a number of ellipses where the distance between foci (the points where you would pin the string if constructing an ellipse in the way suggested in Task 8.4.6) has been varied. The smaller the eccentricity of the ellipse, the smaller the distance between the foci, and the closer it is to a circle.

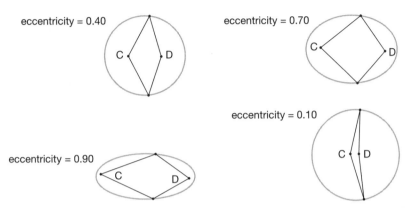

Figure 8.4d

You may be aware that the orbits of planets about the Sun are elliptical with the Sun at a focus of the ellipse. The eccentricities of the orbits of the planets are given in the table below.

	e		e
Mercury	0.206	Saturn	0.056
Venus	0.007	Uranus	0.047
Earth	0.017	Neptune	0.009
Mars	0.093	Pluto	0.25
Jupiter	0.049		

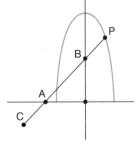

Most of the planets have tiny eccentricities which make the orbit almost indistinguishable from a circle. Ellipses arise in other strange ways too, as Figure 8.4e illustrates.

Figure 8.4e

For thousands of years it was generally believed that the Earth was at the centre of the universe and that the planets of the solar system rotated about this fixed point. Figure 8.4f shows the Earth at the centre of the solar system with all other celestial bodies in circular orbit with the Moon (Lunae) nearest the Earth followed by Mercury and then Venus. This model of planetary motion was carefully argued by Ptolemy who lived *circa* AD 85–165.

In his text, the 'Almagest', Ptolemy synthesised the major results of Greek astronomy and in particular drew on the major works of Hipparchus (190–120 BCE). Ptolemy was able to give a convincing model of the geocentric

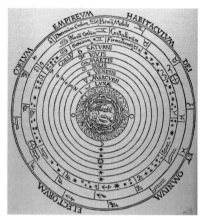

Figure 8.4f

universe that was able to account for the observed motion of the planets. This is not as easy as perhaps might first be thought: prior to Hipparchus it was thought that the planets moved in circular orbits with constant speed with the Earth at the

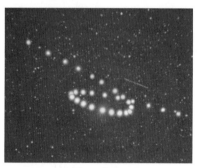

Figure 8.4g

centre of the system – the view of Aristotle (384–322 BCE). However, observations showed that this could not be the case as it did not account for the retrograde motion of the planets. This is when the observed motion of a planet does not continue in the same direction across the sky from night to night. Instead of continuing to move anticlockwise, planets can be seen to reverse direction for a short period before reversing direction again and continuing to move in an anticlockwise direction (Figure 8.4g).

In Ptolemy's model of planetary motion, retrograde motion was accounted for by the introduction of epicycles. This allowed for the planet to orbit around a point that in turn orbited the Earth around a circle. This meant that the constant speed of the planet became non-constant when viewed against the background stars from Earth, and indeed its path included the required retrograde motion. The interactive file '**8m Retrograde1**' represents a Ptolemaic planetary system.

Observations of the planets highlighted problems with the model of Ptolemy. For example, if a planet moved around its epicycle its distance from the Earth would vary and consequently its apparent observed size should also vary. However, observations showed that this did not happen and suggested that, in fact, the geocentric model must be incorrect. However, Ptolemy's model was thought to be correct and used for many centuries until the sun-centred (heliocentric) model of Nicholas Copernicus was accepted (see Figure 8.4h).

The retrograde motion of planets could now be accounted for by the motion of the planets, relative to a

Figure 8.4h

moving Earth. In his major work, *De revolutionibus orbitium coelestium*, Copernicus set out his ideas (although this was only published in the year of his death, 1543). However, at this time Copernicus also needed to include epicycles in his model, to account, for example, for the fact that his model was still based on the assumption that the planets moved on circular paths at constant speed.

The interactive file '**8n Retrograde2**' models the Copernican planetary system. The position of a planet as viewed from the Earth is shown by projecting a ray onto a distant circle representing the distant 'fixed' stars.

Reflection 8.4

What was your response to the multiple ways of constructing parabolas, ellipses and circles?

8.5 PEDAGOGIC PERSPECTIVES

This chapter has focused on developing alternative visualisations and representations via geometry. Graphical understanding has underpinned much of the chapter. The chapter has carefully repositioned geometry back to its important place in allowing you to make sense of a range of mathematical ideas in much the way used prior to Descartes' development of graphical understanding using co-ordinates. This alternative understanding has allowed you to make connections within mathematics and with other areas of study, for example, astronomy.

Research with primary school mathematics teachers (Askew et al., 1997) has shown that learners who are in the presence of teachers who are aware of and make connections between otherwise apparently disparate topics learn most effectively and most richly. Encouraging learners to engage in mental activity (imagining and expressing, conjecturing and convincing) appears to be effective for promoting understanding.

In order to appreciate graphs it is necessary to have a multiple perspective, to see them as at the same time a static object (a curve), a set of points (those points that satisfy some condition), and as the path of a point moving according to some conditions. The conditions may be specified in terms of a restriction on the co-ordinates (an equation or perhaps an inequality), and may be specified in terms of some more geometrical relationship (the ratio of distances from …). Every equation (or inequality) determines a graph (or a region), and every graph can be thought of as a locus: the locus of points for which the x and y co-ordinates satisfy the equation (inequality).

Trigonometry is a rich domain of relationships that arise from the need to be able to measure angles in different ways, as ratios. The whole subject actually makes sense only because Thales' theorem guarantees that the trigonometric ratios associated with a given angle are invariant under choice of the sizes of triangles using that angle.

Locus and trigonometry are rich domains for exploring with body movements. For examples, learners can walk in such a way that they are always a constant ratio of distances from two fixed points. Body movements support mental imagery, as does seeing images generated by dynamic geometry software. It is important not to become reliant on either acting things out or using dynamic geometry software. The most important human power is the use of mental imagery, and learning to express what is imagined in mathematical ways and making sense of mathematical phenomena.

As mentioned in Chapter 4, Jerome Bruner (1986) suggested three different modes of presentation or encounter with phenomena: enactive (bodily), iconic (imagery including diagrams) and symbolic (words and symbols). He proposed that almost any topic can be taught to almost any learner if it is suitably set in a context through which the learner can experience at least something of these three modes. Certainly, rushing learners into symbols before they have become fluent with expressing their thinking is more likely to generate frustration and disaffection than real progress. That is why some teachers find it helpful to think in terms of getting learners doing something, talking about what they are doing and, when that talking becomes fluent, provoking them into making records (diagrams, words, symbols). Of course, attempting to record often helps clarify meaning: it is not a one-way process but more of an endless spiral or circular return. These ideas are elaborated in Chapter 14 under the headings of 'Enactive–Iconic–Symbolic' and 'Do–Talk–Record'.

Reflection 8.5

You have now finished Block 2. Which parts of Block 2 do you need to do more work on?

What references to the history of mathematical ideas in general, and geometrical ideas in particular, might you and/or the learners with whom you work be interested in following up on the web?

What have you learned about applications of geometry?

Introduction to Block 3

This block is concerned with geometrical thinking in terms of transformations which map one object to another, and which map the whole of a space to itself.

Chapter 9 uses the idea of invariance to distinguish between and to characterise different geometrical transformations. The chapter develops into consideration of combinations of transformations, and transformations in three dimensions.

Chapter 10 is the third chapter in the book to focus on points of view and language, this time in the context of geometrical transformations. The chapter starts with paper-folding and tessellations. It leads on to the notion of a *dual*, which is a lovely example of a transformation that does not preserve congruence. It takes the idea of a dual into three dimensions, and considers applications.

Chapter 11 draws on some of the ideas in Chapter 7 and approaches them from a more explicitly 'transformation' focus. Reasoning using transformations is compared with other ways of reasoning, moving between Euclidean, measurement and transformation-based perspectives.

Chapter 12 starts by considering transformation as dynamic rather than static. It looks at composite rotations and reflections, and how attention in transformation geometry is drawn to the intermediate positions between start and finish of a transformation.

Each chapter ends with a review of pedagogic issues raised.

9 Transformations and Invariants

This chapter uses the fundamentally important idea of invariance to characterise and to distinguish between different geometrical transformations.

The first section is about three transformations that keep shapes the same size. These are known as *isometries*, from the Greek *iso-* meaning equal, and *-metry* meaning measure. In contrast, the second section considers some transformations which preserve shape but not necessarily size.

The third section focuses on combinations of transformations and what they leave invariant. Section four looks at transformations in three dimensions and the chapter ends with a pedagogic review in section five.

Task 9.0.1 Tiling

Figure 9.0 indicates a part of a floor tiling. What do you have to do to convince yourself that the tiling can be extended indefinitely?

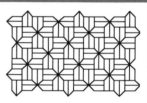

Comment

One approach is to pick out a portion and to find it re-occurring somewhere nearby, then to see if that re-occurrence is repeated consistently in a manner that permits indefinite repetitions. You can focus on a single trapezium, or on combinations such as a pentagon (two tiles) or a cross (eight tiles).

Figure 9.0

9.1 FLIPPING, TURNING AND SLIDING

This section is about three transformations that keep shapes the same size, known as *isometries*.

Task 9.1.1 Check and Tartan

What is the difference between CHECK and TARTAN?

Comment

Did you think about the meanings of the words *check* and *tartan* as differences between uniform division into squares and non-uniform division into rectangles? An alternative interpretation is to look at the words themselves. Written in a suitable font, the word CHECK is invariant (more or less) under reflection in a horizontal mirror, where as TARTAN is not.

Like many jokes, this one makes use of a switch of interpretation, of focus of attention. It intentionally attracts attention to the meaning of the words, then switches attention to the words themselves. A good deal of teaching of mathematics has a similar quality, in that attention needs to be attracted to mathematically relevant features, relationships and properties in order to make sense.

Neither *check* nor *tartan* are invariant under other mirror reflections, whereas words like MOW are (more or less) invariant under a rotation through 180°, and OXO is invariant under two different mirror reflections *and* a rotation. Searching for other words or letter combinations with these properties is a good way to invite learners to think about transformations because they are using mental imagery to test possibilities.

When an object is invariant under a transformation, it is usually described as having symmetry. Symmetries are transformations of an object to itself. One of the breakthroughs in modern mathematics came as a result of switching attention from symmetries of objects to invariants of transformations of those objects. It turns out to be even more fruitful to study transformations between objects, and even to study the effects of transformations on the whole space, not just on individual objects. For example, the relation 'is congruent to' is an assertion of the existence of a particular kind of transformation (an *isometry* or distance-preserving transformation) which takes any one shape to another; the relation 'is similar to' (in the mathematical sense) is an assertion of a transformation which preserves shape but not necessarily size. These relationships lie hidden at the heart of Euclidean geometry, but it is only in the last 200 years that people have thought explicitly in terms of transformations.

Most people are familiar with mirror reflections, at least when presented in standard position as in Figure 9.1a. But as an example of a mirror reflection this diagram has some notable shortcomings: there are features of reflection which are absent and hence potentially confusing.

Figure 9.1a

Task 9.1.2 Reflections

What dimensions of possible variation are you aware of which bring out some of the nuances of mirror reflections?

Comment
Did you think of changing the position of the mirror? Although conceptually it makes no difference to the mathematics, psychologically it makes a big difference. Many learners find it hard to recognise a mirror reflection when the mirror is neither vertical nor horizontal.

Did you think of changing the shape so that the effects of mirror imaging are more distinguishable? The square has mirror symmetries of its own, so it is not immediately clear which points of the square are mapping to which points of the image.

Of the three transformations, it is easiest to pick out translations (sliding without rotating), so that means you can concentrate on distinguishing rotations (which preserve orientation) and reflections, which reverse orientation.

In the next task, you are asked to use rotation and reflection.

Task 9.1.3 Distinguishing Reflections, Rotations and Translations

In Figure 9.1b there are two sets of eight shapes, quadrilaterals and pentagons. Within each set, each shape can be transformed into any other by use of a rotation or a reflection. Watch how you identify the transformation as you work on various pairs. What experiential differences are there in working with the pentagons rather than the quadrilaterals?

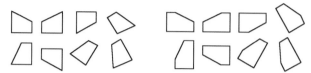

Figure 9.1b

Comment

The aim is to transform one object so that it lies properly on top of the other. As you focus on a pair can you 'feel' the transformation?

Start with the first object in the first row and decide what transformation takes it to each of the others. Some people experience a kinaesthetic sense as they imagine one object rotating and/or reflecting to end up on top of another. You will have found that for the bottom-left pair in each row, rotating one about its 'centre' and then translating (sliding) to fit on top of the other was mentally easier than finding a point around which to rotate the one into the other. Similarly, for the last two in each row, reflecting in a diagonal mirror orients the shape correctly, and then translating (sliding) gets one on top of the other.

Task 9.1.3 invoked an important mathematical idea, namely, the composition of transformations: following one with another. Perhaps there are other transformations which can be achieved which are not one of the three? For example, if you rotate a shape around one point through some angle, and then about another point through some other angle, what is the result? It is not obvious that the result is a rotation, about yet another point, through the sum of the two angles. What happens if you combine rotations and reflections?

It turns out that every isometry is a combination of reflections and, furthermore, you never need more than three mirrors to reproduce any isometry. What also emerges is that there is one other isometry, a less familiar one, which is a combination of three mirrors, two parallel and one perpendicular to the others.

Task 9.1.4 Combining Isometries

Figure 9.1c shows two pairs of mirror lines. In each case, what is the effect of combining reflection in one with reflection in the other? Does the order in which you use the mirrors matter?

Figure 9.1c

Comment

At this point, you can experiment with Interactive files '9a Transformation One' and '9b Transformation Two', using either of two modes: you can expose the construction for the transformation and explore its effects, or you can keep the construction hidden and work out what the transformation is using test items such as a point, or a segment, or a circle, or a quadrilateral.

Reflection in parallel mirrors has the same effect as a translation. So any translation can be thought of as a sequence of two reflections. But there is considerable choice available: if the translation is specified, the first mirror can be placed anywhere (but its direction is already determined). Something similar happens for rotations, as the next task demonstrates.

Task 9.1.5 Two Mirrors to Make a Rotation

Imagine a point and a rotation about that point through a specified angle. What choices have you got for mirror lines so that the combined reflections produce the same effect as the rotation? What happens if you reverse the order of using those mirrors?

What is the same and what is different about having the mirrors parallel, and having them on intersecting lines?

Comment
Some experimenting mentally, supported by a diagram or by using interactive files, shows that one mirror can be chosen freely as long as it goes through the centre of rotation (or is perpendicular to the direction of translation). The second mirror is then determined: it passes through the centre of rotation and makes an angle of half the desired rotation angle with the first mirror.

A translation can be seen as a special case of a rotation, about a point which is infinitely far away!

Mathematicians find it convenient to avoid having to add in special cases, such as 'what happens when the mirrors are parallel', so translations are seen as special cases of rotations, just as a square is a special case of a rectangle, rhombus and parallelogram.

A question which immediately occurs to a mathematician is to find out what happens when you perform two rotations in succession. Seeing rotations as made up of two reflections is a very useful way to analyse the situation as the next task shows.

Task 9.1.6 Two Rotations

Imagine a rotation about the point P through an angle of α (read 'alpha') anticlockwise. Imagine following that with a rotation about the point Q through an angle of β (read 'beta'), again anticlockwise. What is the overall effect on P and on Q? What is the overall effect on other points?

Suggestion: present the first rotation as the composition of two reflections, the second being the line through PQ. Present the second rotation as a composition of two reflections, the first being the line through PQ.

Comment
By convention, angles are measured anticlockwise. Notice that no diagram is given – you are encouraged to draw your own!

Following the suggestion, the overall effect is a sequence of four reflections. But the middle two reflections are in the same mirror line and so overall have no effect. The full sequence is really the same as the composition of two reflections and hence a rotation. You might want to check out the effects when one or both are translations.

Tangrams were mentioned in Chapter 1 as a source of geometrical problems which can be used to work on geometrical thinking. While working on tangram puzzles, attention can be drawn to transformations of individual pieces, but without explicit attention these aspects are likely to be overlooked by learners.

So far, the only use of three reflections has been to show that the composition of two rotations is itself a rotation. There is one more type of transformation which was mentioned earlier that requires three reflections to specify it.

Task 9.1.7 Glide Reflections

Imagine a shape. Translate it in some direction along a line (this can be done with two reflections). Then reflect the result in a mirror parallel to the translation line. Draw several examples for yourself and try to find a way to produce the same effect with just two reflections.

Comment

The translation direction is the direction of *glide*; that the line parallel to the glide direction becomes the mirror for the reflection; hence the transformation is known as a *glide reflection*. Two reflections are not enough to achieve the same effect: some examples can be found in the Interactive file '9c **Transformation Three**'. Combining three reflections is always either a reflection, a reflection combined with a rotation or a glide reflection. Using four reflections does not introduce anything that could not have been achieved more simply using three or fewer.

In thinking about transformations so far you have been thinking mainly in two dimensions, that is, transformations of the plane. In fact, transformations are also applicable in three (and more) dimensions. Translations are reasonably easy in three dimensions, rotations occur about a line rather than a point, while reflections require a mirror plane rather than a mirror line. It is unfortunate that school curricula do not provide interested pupils with the opportunity to explore more geometry including transformations in three dimensions.

Reflection 9.1

How do you find yourself moving from the verbal to physical, mental and symbolic representations? What are some of the difficulties in trying to support your learners in getting to grips with the idea of transformations?

9.2 SQUEEZING, STRETCHING, SCALING AND SHEARING

The transformations encountered so far have been isometries: distance between points has remained unchanged, so a figure and its image are congruent. This section considers some transformations in which distance changes.

Similarities

Scaling uniformly in all directions (also called an enlargement, even when the image is smaller) is an important transformation at the heart of scale drawings for engineering and artistic design, whether in two or in three dimensions. For example, if you take a shape, and a point C which will be the *centre of enlargement*, and if for each point P on the shape you draw a line CP and then mark a point which is, say, two-thirds of the way from C to P, then you will get a new shape, which will be two-thirds the size of the original.

Clearly lengths are not preserved, but of course the ratio of corresponding lengths is preserved. There are two somewhat surprising facts about uniform scalings.

Task 9.2.1 Uniform Scalings

In Figure 9.2a two copies of the same shape are shown, at different scales but no rotation or reflection involved. Imagine joining corresponding points with straight lines: what do you think might happen (make a conjecture before you try it!). Why must that always happen?

Suppose you scale a drawing using a *centre of enlargement* C, and someone else scales by the same factor using a different centre of enlargement D. What will be the same, and what different about the two images?

Figure 9.2a

Use the interactive file '**9d Transformation Four**' to explore uniform scaling.

Comment
All the lines must meet at a common point, the centre of enlargement. The reason is due to Thales' theorem (corresponding edges in the two shapes are parallel). Using different centres of enlargement produces the same scaled object but positioned differently. Even though it seems as though points closer to the centre do not get moved as far as points more distant from the centre, what *is* preserved is the ratio of the distances (the scale factor).

The property of all the lines joining points to their corresponding images meeting at a common point actually characterises uniform scalings: if the lines all meet at a point, then Thales' theorem tells you that the corresponding ratios are all equal.

With an enlargement you always get a similar figure. All the sides of the image are scaled by the same scale factor and all the angles of the image are congruent with the angles of the original. Parallelism is preserved, and a line and its image are always parallel. When the scale factor is negative each point goes the other side of the origin, so the orientation (clockwise/anticlockwise) changes. Thus orientation is not invariant under all uniform scalings.

Two shapes (think triangles) can be similar without one being a simple scaling of the other, because one might also be rotated through some non-zero angle. Combining uniform scalings with the isometries gives the collection of all transformations which preserve similarity of figures. Notice that although lengths are not preserved, ratios of lengths are preserved, as are angles.

Stretches

An extension of uniform scaling is to stretch or scale along one straight line by one factor f_1, and along another straight line by a second factor f_2. Because it is a bit complicated, you might wish to consider using a scale factor of 1 for one of the directions (or use the interactive file '**9e One-axis Scaling**') for your first read through the following.

The circle in Figure 9.2b has been inscribed in a square with three additional diagonal line segments added. Imagine what would happen if each point was transformed into a point whose y-co-ordinate was doubled and whose x-co-ordinate was halved.

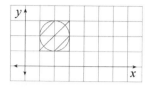

Figure 9.2b

Drawing the image and tracking what happens to the various line segments and the circle suggests that lines remain lines. Where lines are incident, the incidence is preserved by the stretching. Parallel lines remain parallel. Ratios of distances along lines is invariant, and area is scaled by the product of the scale factors (think Cavalieri: scaling in one direction stretches all segments parallel to that line by that factor so the area is also scaled). The circle changes to an ellipse, and its area is multiplied by the same scale factor.

Formal analysis of such transformations is not appropriate here, but there is an interactive file '**9f Non-uniform Scalings**' to give you a taste of what sorts of things happen.

As with the congruence transformations, it is possible to present a two-way stretch as a combination of two one-way stretches, leading to questions about what happens if you reverse the order of two one-way stretches, and what happens if you have two stretches along the same line. Stretches can also be combined with congruences and similarities.

When the two stretch factors are the same then the result is the same as an enlargement (or uniform scaling transformation) where the centre for the enlargement is the point where the invariant lines intersect and the scale factor is the same as the stretch factor. As the scale factor works in all directions, all lengths are changed by the same factor and the area is changed by the square of the scale factor.

Shears

Thinking back to Figure 9.2b, imagine what would happen if every point moved parallel to the x-axis a distance equal to half its distance from the x-axis.

Note that points move to the right by an amount determined by their height above the x-axis.

Task 9.2.2 Shearing

Imagine what would happen to each of the following elements in Figure 9.2b when sheared *before* you do any drawing! It takes a little thinking about, but it can be done.

Which lengths change and which stay the same? What is the ratio of the changed length to the original length in each case?

Which angles change? Which incidences change? Which pairs of parallel lines are no longer parallel? What happens to the area of the square?

Comment
This transformation is an example of a shear. Each point on the x-axis remains fixed while all other points move to the right (the factor of half could of course be varied, and the choice of the x-axis could be any line). The interactive file '**9g Shear**' shows what happens under a shear, by using an inclined line to indicate how much each point moves according to its distance from the x-axis.

A shear is a curious transformation. One line is invariant, and the further any point is from that line, the further it moves. Points above the line move one way, points below the line move the other way. (If P is 3 units above the line, and the scale factor is 2, then it moves 3 × 2 units.)

Parallel lines stay parallel; angle size changes; the height of a figure (perpendicular to the invariant line) stays the same; the area stays the same; the perimeter changes.

There are potential confusions between shears and stretches. For example, four equal rods hinged at the ends can form a square, but will also flop over to form a rhombus as shown in Figure 9.2c.

Figure 9.2c

What transformations are involved? Shearing could be used to achieve the desired angle, but then the height would have stayed the same, and the tilted rods would therefore have become longer thereby following up with a scaling down in the direction of the tilted rod to re-establish the equal lengths. Thus, a 'sideways flop' is a shear combined with a one-way stretch.

Reflection 9.2

Did you stop, imagine, draw your own diagrams and invent your own exploratory tasks. Or were you content to follow, and then skip over, the exposition?

9.3 INVARIANTS AND COLLECTIONS

One of the breakthroughs in modern mathematics was to characterise transformations in terms of what they leave invariant, rather than thinking about what they change. The features considered include length, angle (including parallel lines and perpendicular lines), incidence, area and ratio of lengths along lines.

This section focuses on collections of transformations and the invariants associated with them.

Composition

Shifting attention from objects being transformed, via invariants, to the transformations themselves led mathematicians to see that composing two or more transformations produces new transformations. It turns out to be very useful to look for collections of transformations that do not produce new ones, that is, which are closed under composition. For example, the isometries are *closed*, that is, every combination of reflection, rotation, translation or glide-reflection produces another one. The collection (or *set*) of similarities is also closed; the collection of isometries, similarities, shears and two-way stretches are closed as well and is called the set of *affine* transformations.

When considering sets of transformations, it is important to remember a very simple transformation, which 'does nothing', that is, which maps every point to itself. This *identity* transformation plays the role that zero plays in addition, and that 1 plays in multiplication: it has no effect. But its presence in the set of transformations is vital for closure. Thus the identity transformation is a special case of a rotation (through angle zero), a translation (zero distance in any specified direction!) and a scaling (scale factor one). It is not a special case of a reflection (because it preserves orientation which reflections do not).

Keeping track when you compose transformations requires notation, and care. You can use $T(P)$ or 'T of P' to mean the image of P under the transformation T. It often makes a difference in which order you perform transformations, so that if T and S denote two transformations, then S o $T(P)$ denotes the transformation formed by applying T to P first, and then S to the result. The little 'o' is read *composition* or *composed with* or simply *of* as in 'S of T' of P.

The next task offers an opportunity to work with this notation by thinking about translations.

Task 9.3.1 Translations on the Line

Let T_a, T_b, and T_c denote three translations along the same line (think number line) by the amounts a, b and c respectively. Using the notation of composition, what would T_b o T_a and T_a o T_b denote? What can be said about T_a o $(T_b$ o $T_c)$ and $(T_a$ o $T_b)$ o T_c?

Comment
In the case of translations along the same number line, the subscripts look like ordinary addition, the composition of translations is commutative (order does not matter) and associative (you can bracket three in a row either way and get the same result.

A bit more interest arises on the line when reflections are included.

Task 9.3.2 Isometries on the Line

Let R_d, R_e and R_f denote reflections of a number line in the points d, e and f respectively. What translation is equivalent to R_e o R_d? Explore commutativity and associativity of reflections.

Comment
Did you apply the general reflections to specific (possible generic) numbers on the number line, like x, or 3?

Did you specialise d and e to numbers, to draw a number line and to try several values? Did you think about the effect of composing two reflections?

Working out $R_d(x)$ takes a little bit of thought (test it on d itself, which should be invariant); applying that same idea to R_e o $R_d(x)$ reveals the translation which as expected is through double the distance between the mirrors. Here order does matter!

Reflections effectively introduce multiplication by $^-1$. In the next task, you will work on scaling more generally.

Task 9.3.3 Similarities on the Line

Denote by S_p, S_q and S_r three scalings of the number line by p, q and r respectively (none of them zero) each leaving 0 invariant. Explore compositions of S_p, S_q and S_r (commutativity, associativity). Compare the effects of S_p o T_a and T_a o S_p, and of S_p o R_d and R_d o S_p.

Comment

Tasks like these provide an opportunity to challenge yourself, to specialise just to the extent necessary so you can keep a grip on what is happening, and then to re-generalise for yourself. As you perhaps anticipated, scalings are a geometrical version of multiplication, and this is reflected in the subscripts, since $T_a \circ S_p(x) = T_{pa}(x)$ while $S_p \circ T_a(x) = p(x + a) = T_{pa} \circ S_p(x)$.

Undoing

There is one further feature of composing transpositions which makes them very powerful in mathematical thinking: when you want to do a transformation in one place, it is often more convenient to do it somewhere else, so you translate (or rotate) to a more convenient place, perform the transformation and then translate or rotate back.

For example, suppose you have a fixed mirror but you want to reflect in a mirror line parallel to it. To find the image of a point P, use a translation T which translates the desired mirror line to the available mirror; perform the reflection R and then translate back again. The undoing or reverse transformation is written T^{-1}. It has the property that $T \circ T^{-1} = T^{-1} \circ T = I$ the identity or do-nothing transformation. In other words, each undoes the other. The reflection can then be written as $T^{-1} \circ R \circ T$ which means do T first, then R, then undo T.

Again, to rotate about a point P by a rotation through an angle θ, first translate the plane so that P is sent to a convenient place such as the origin, then perform the rotation R_θ, and then translate back again.

The collection of all reflections can be generated by using reflections in mirror lines through a single point, together with translations. The collection of rotations can be generated by using rotations about a single point, and translations.

Tiling

Figure 9.0 at the beginning of the chapter showed part of a tiling. It has been repeated in Figure 9.3a with some selected tiles shaded in.

Figure 9.3a

Task 9.3.4 Transformations in Tilings

Using the shaded tiles, locate pairs which are related by as many of the transformations discussed in section one as possible.

Comment

The transformations all have to be isometries because the tiles are all congruent. Remind yourself to look out for patterns in wallpaper, fabric, and tiles in which individual parts are repeated, and look for the transformations involved.

The words tiling and tessellating are often used synonymously. Technically, to tessellate means to cover the plane using only one shape, while to tile means to cover the plane with a repeating pattern that may use more than one shape. Squared paper and isometric (triangular) paper are simple examples of tessellations. Testing a tile to see whether it tessellates can be trickier than you might expect.

Task 9.3.5 Tiling and Tessellating

Find which of the shapes in Figure 9.3b can be used to tessellate a plane. Describe each tessellation in terms of the transformations used.

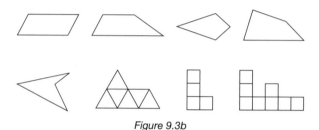

Figure 9.3b

Comment

All of the shapes shown will tessellate, although some are not immediately obvious.

Because the angles of a quadrilateral and the angles at a point both add up to 360°, it is reasonable to think in a tessellation that at each vertex, all four corners of the original quadrilateral must occur.

You may be surprised that any quadrilateral will tessellate, even non-convex ones! If you are not convinced, you may wish to explore the interactive file '**9h Quad Tess**'.

Each manipulation of an object, whether by dragging something to change a shape or operating a transformation on an object, can contribute to your sense of what is possible, but is likely to have an effect only if you reflect enough to make sense of what you are seeing. When using any materials, electronic or physical, it is easy to become caught up in the manipulations without appreciating what you are doing and without making mathematical sense. In this case, you will have discovered that any and every quadrilateral will tessellate the plane, except those whose edges actually cross. You may have noticed that '**9h Quad Tess**' uses rotations about the mid-points of the edges of the quadrilateral. Did you think about which of the angles of the quadrilateral show up at each vertex?

Thinking in terms of what is left invariant by a transformation or a collection of transformations, and in terms of how much freedom a collection has to allow the construction of a mapping between two pre-assigned objects gives complementary ways of thinking. Composing transformations, and treating two objects as *equivalent* if each can be mapped to the other by some transformation in a predetermined set of transformations, lie behind the development of powerful mathematics of the twentieth century.

Reflection 9.3

What has changed in how you look at patterns and how you think about transformations?

9.4 THREE DIMENSIONS

Embedded as we are in three physical dimensions, we are not always as aware of trans-formations in three dimensions, perhaps because we cannot 'look down' and see things happening. But they are all around us. This section looks at transformations in three dimensions.

Task 9.4.1 Using a Projector Screen

What happens with the screen and a projector if the screen is at a slight angle sideways, vertically, or both? In other words, what happens to the image if you tilt the screen a little?

What happens to parallel lines; to intersections of lines; to right-angles?

Comment
The result is known as the 'key-hole' effect. Lines still map to lines, but the relative lengths change. If the top of the screen is closer to the projector than the bottom then the length along the top will be smaller than the length on the bottom of the image. The image of a rectangle will be a trapezium with the longer parallel side at the bottom. Horizontal lines will stay parallel, but other lines will not, and angles will change.

Another way to experience the same phenomenon is to cut a square hole (about 3 cm on the side) out of a piece of stiff card. Then draw any convex quadrilateral on a piece of paper. Orient the card relative to your eye so that when you look through the hole, the sides of the hole match the sides of the quadrilateral. The square is being projected onto the quadrilateral from a point in your eye.

The projector screen is performing a projection from one plane onto another. Such a transformation is called a projective transformation. Although it seems that a projective transformation leaves little if anything invariant, a breakthrough came when Jean Victor Poncelet (1788-1867) and Joseph Gergonne (1771–1859) inde-pendently created *projective geometry*. Poncelet did so while imprisoned during the Napoleonic wars, having nothing else to do in his cell. The essence of projective geometry is an invariant developed in the next sub-section.

Projections

In Chapters 5 and 8 you met various projections that were used to achieve apparent three-dimensionality of drawings on paper and to map surfaces.

Task 9.4.2 Projections

What is the same and what is different about the two transfor-mations illustrated in Figure 9.4a, projecting from a point and orthogonal projection?

Comment
Neither of these types of projection preserves lengths. However, Thales' theorem applies to the orthogonal

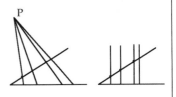

Figure 9.4a

---

projection to show that the ratio of lengths along a line is preserved. For the point projection, even ratios are not preserved. But amazingly, something is preserved! How anyone noticed it is a mystery.

Using the sine relationship in the various triangles (Figure 9.4b) it is possible to show that

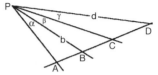

$$\frac{AB}{BC} \times \frac{DC}{AD}$$

Figure 9.4b

is independent of the angles and the distances from the point P and depends only on the four lines through P; it is also independent of which transversal line cuts those four lines. Thus this ratio of ratios (product of two ratios) is an invariant for projection from a point.

This result has implications in three dimensions and beyond. For example, it turns out that if you perform a point projection of a triangle with points on its sides onto another plane as in Figure 9.4c, then the product round the triangle,

$$\frac{AP}{PB} \times \frac{BQ}{QC} \times \frac{CR}{RA}$$

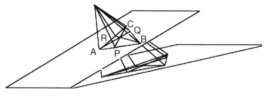

is invariant under the projection. In fact, when this product is one, it can be proved that the lines AQ, BR, and CP all coincide at a single point, when the product is

Figure 9.4c

-1, and, taking directions into account, the three points P, Q and R all lie on one straight line. These two are therefore projective invariants.

By comparison, the three-dimensional features of orthogonal projections are much more straightforward.

Task 9.4.3 Orthogonal Projections

Use Figure 9.4d to show that the ratio of lengths of a segment AB and its image depends only on the angle between the planes, so the ratio of lengths along a line is invariant under orthogonal projection.

Extend this, using Cavalieri's principle, to show that ratios of areas is also preserved under orthogonal projection.

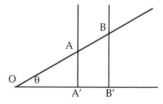

Figure 9.4d

Comment

Did you use Thale's theorem, or did you calculate the lengths of OA and OB and their images using cos θ? Since the ratio of lengths of AB to its image A'B' is cos θ, the ratio of lengths of two segments AB:CD on the same line will be the same as the corresponding ratio of the images. To compare areas you need to think in terms of a region on one plane being projected onto another plane. By thinking of the area as made up of line segments at right angles to the line of intersection of the two planes, each is scaled by the cosine of the angle between the planes, so the areas are also in the same ratio. Thus the ratio of two areas is the same as the ratio of the projections of those areas.

Conics

Apollonius, mentioned in Chapter 8 in connection with circles, is best known for his work on conics: ellipses, parabolas and hyperbolas. In Chapter 8, ellipses and parabolas were introduced as the loci of a point moving under different constraints. The name *conics* comes from the origin of the shapes as projections of a circle (hence, a cone).

Task 9.4.4 Projections of a Circle

Imagine a circle, and a point O not on the plane of the circle. Join each point of the circle to the point O by a line (infinite in both directions). The result is a full cone (extending in both directions). Imagine cutting this cone with three different planes. (See Figure 9.4e.)

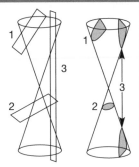

The first plane (1) should be parallel to the edge of the cone; what shape would you expect the intersection of the cone and the plane to be?

The second plane (2) should be at a different angle so that the plane makes a closed curve with the cone; what shape would you expect the intersection of the cone and the plane to be? In this case what would happen if the plane was parallel to the base?

Figure 9.4e

The third plane (3) should be at a different angle so that the plane cuts the cone twice; what shape would you expect the intersection of the cone and the plane to be in this case?

If possible, imagine the cutting plane slowly rotating from position 2 to position 3 via position 1 so that you see the transitions between the various sections.

Comment

It is worth struggling a bit to form an image, and perhaps helpful to make (part of) a cone out of a sheet of paper so as to support your imagining. Once you look at someone else's pictures, you have to work to 'see' how the picture displays what was being described.

  The parabola is distinctive because as the cutting plane changes its angle the ellipse becomes more and more extreme until suddenly it is a parabola when the cutting edge is parallel to the edge of the case and then, just as suddenly, an extreme hyperbola.

Each such curve formed by cutting a cone with a plane is the projection of a circle onto that plane. The curves you can achieve are either parabolas, ellipses, (circles are special cases), or hyperbolas.

  Thinking in terms of projections makes it straightforward to ask questions such as: if you form a cone from a parabola (or an ellipse, or a hyperbola), what shapes can you get by cutting the cone with different planes? Since the parabola is the projection of a circle, you still only get projections of circles, which are again conic sections.

Surfaces

You have already been invited to think about an ellipse as a two-way stretch of a circle. Now imagine rotating an ellipse about one of its axes to produce a surface. One of its cross-sections is circular, while the others are elliptical. Other surfaces can be generated by rotating conic sections.

Task 9.4.5 Imagining Surfaces

Imagine a parabola. Imagine rotating it around its axis so that it has a circular section. Now imagine all the circles being stretched in the same direction and with the same stretch factor to make ellipses. This is a *paraboloid* with elliptical cross-sections.

Having combined a parabola with an ellipse, perhaps it can be combined with a hyperbola. Imagine a parabola with its vertex pointing to the left and moving so that its vertex is always touching a hyperbola. The solid shape swept out by this moving parabola is called a *hyperbolic paraboloid*.

Imagine a hyperbola. The hyperbola has two axes, one intersects the hyperbola, the other does not. Rotate the hyperbola around the axis that *does not* intersect it. Stretch all the resulting circles so that they form ellipses. This is a *hyperboloid of one sheet*. The reference to 'one sheet' draws attention to the fact that the surface is one connected piece, in contrast to what happens if you use the other axis.

Imagine the hyperbola again. This time rotate it around the axis that *does* intersect it. Stretch all the resulting circles so that they form ellipses. This is a *hyperboloid of two sheets*.

Comment

If at all possible, try this task with a partner.

These imaginings are by no means simple, but it is worth persevering before looking at the pictures below.

When you look at the pictures, consider how the imagining description fits with what you see.

These shapes are probably unfamiliar. It is easy enough to think of rotating a shape and getting a circular cross-section, a little harder to imagine an elliptical cross-section. The hardest to imagine is the hyperbolic paraboloid. It is interesting because it forms a 'saddle' because it is concave and convex at the same time – to see the saddle look at Figure 9.4f from the side.

Your images may, or may not, bear some similarity with these pictures. Other pictures can be found on the web using a search engine.

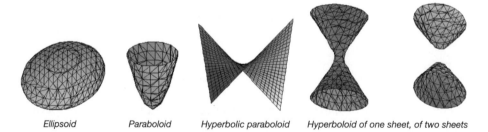

| Ellipsoid | Paraboloid | Hyperbolic paraboloid | Hyperboloid of one sheet, of two sheets |

Figure 9.4f

The hyperbolic paraboloid is interesting because it can be made with straight lines. This is why it is sometimes used as a roof. In the next task you can visualise the straight line generating the shape, then make one with curve stitching.

Task 9.4.6 Making (part of) a Hyperbolic Paraboloid

Imagine holding a metre ruler just in front of your eyes so that it makes an angle of about 30° to the horizontal. Now slowly move it away from your eyes and as you move it steadily rotate it clockwise so that at first it is horizontal and then when it is at arms length it makes an angle of 30° the other way (that is, you have rotated it through 60°). Now imagine doing this again and think about the surface that the ruler would sweep out. Does it help to use a physical ruler?

Comment
When imagining things, it is often hard to keep hold of all the features as you focus on something particular. It is better to let go of other features but to refresh their presence every so often, rather than trying to retain them all as in a picture. When you have a picture to look at, you ignore many features in order to focus on others. The difference is that you know the other features are available when you choose to refocus, whereas with imagery those features might have to be reconstructed.

To make a hyperbolic paraboloid get four strips of wood (or very stiff cardboard) about 30 cm long with holes bored every 3 cm along the strips (Figure 9.4g). Firmly tie the corners together to make a rhombus, then bend the rhombus at a diagonal and support it as in the diagram with a support 15 cm high. Next, thread a long thread (about 4 m) through the holes (or round the tacks) from one side of the frame to the other firmly fixing the thread at each end as in the figure. Repeat with another thread using the holes on the other two sides. Finally, remove the support and smile.

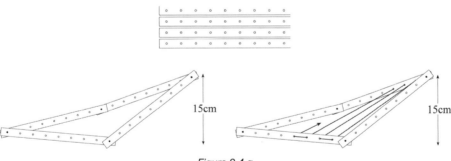

Figure 9.4.g

You will now have a curved figure generated by straight lines. Position it so that it balances on the two low points and imagine these being set in firm foundations. Look for the saddle point in the centre of your model.

A hyperboloid of one sheet can be made by drilling holes around two circles, one on each of two sheets of wood or very stiff card, and then stitching the two circles together but holding them about 30 cm or more apart. The result is a cylinder, and when one circle is rotated with respect to the other, the result is a hyperboloid of one sheet with circular cross-section.

Stretches, Shears and More

Just as one-way stretches in two dimensions can be composed to produce two-way stretches, so in three dimensions you can have three-way stretches. Imagine a sphere (the three-dimensional analogue of a circle). Imagine stretching each of the co-ordinate axes by a different factor, say a, b and c respectively. The sphere will be stretched differently in each direction to form an ellipsoid: a shape which is elliptical in each cross-section. It looks like a rugby ball. Then as in the area of an ellipse in the two-dimensional case, the volume of any shape will be scaled by a factor of abc. The volume of a sphere is $4\pi r^3/3$ and, more generally, the volume of a three-way-stretched sphere, an ellipsoid, with semi-axes a, b and c, is $4\pi abc/3$.

Shearing in three dimensions involves sliding planes sideways proportional to their distance from an axis plane, and shears can be composed. Furthermore, shears can be composed with stretches. Analysis of all the possibilities is most easily accomplished using matrices and linear algebra.

Reflection 9.4

What role did the practical examples play for you in thinking about projections, including using a sheet of paper to make a cone, and using a ruler to imagine the hyperbolic paraboloid? What happened when you switched from imagining to interpreting the diagrams provided?

9.5 PEDAGOGIC PERSPECTIVES

In this chapter you have considered transformations as actions upon objects and as actions upon the whole plane (and even the whole of three-dimensional space). Thinking about transformations as objects leads in several directions: to composing them to form classes of transformations; to seeking invariants which characterise various transformations (length, angle, area, parallelism); to seeing relationships between shapes (congruence, similarity, area preservation) as determined by the collection of transformations permitted (isometries, similarities, shears). Similar work can be done with orthogonal projections and point projections.

There has been an alternation between task-driven experiences and exposition in order for you to reflect upon when one or the other might be more appropriate when working with learners. A steady diet of all one or all the other can become numbing, as perhaps you have noticed. Learners do not always know what to do with tasks (they need to re-learn to become independent of explicit and totally clear instructions in order to re-develop a problem-solving approach to situations). There is also a risk of learners getting stuck and being unable to complete the task; various strategies are helpful in overcoming this such as specialising in order to re-generalise; finding a physical presentation from which to get a sense of what goes on; seeking relationships, and so on.

Being text-based, the tasks presented here are often in the form 'imagine ... '. Increasingly you have been expected to draw your own diagrams and to work mentally on those diagrams. There have been suggestions to use interactive files, and opportunities to cut out shapes and to move them about. Working with physical objects can focus your attention on achieving results without actually developing imagining powers. Prompting learners simply to look at physical objects and diagrams and to imagine parts moving but preserving the structure, or superimposing further elements makes use of learners' powers and helps those powers to develop. There is more on these ideas in Chapters 13 and 14.

Visual–Aural Preferences

It is important to remember that some people find it easier to respond to mental imagery invitations than others. This is because some people 'see' in their minds, or at least respond readily to the language of 'seeing', while others are more auditory in focus, hearing words and finding them come to mind more readily. Everyone has the powers necessary which can be developed, but it may take time. Getting learners to discuss their ways of thinking can be very effective in opening up possibilities that they have previously overlooked.

There is a wider dimension to the visual–aural preferences which some learners present. Imagine the mismatch in the classroom where some pupils and a teacher might be discussing ways of working:

I think in pictures and diagrams	I teach with words
I want the whole picture	I want you to look at the details
I want to synthesise	I want you to analyse
I want to know what	It is important that you know why
I make intuitive leaps	Every small step is important
I like challenging puzzles	I want you to master the basic facts

Learners do not usually articulate these differences, indeed they may not be aware of them, but in most classes some learners think differently from others and from the teacher. One coping method is to set tasks that can be approached in different ways, then notice, acknowledge and celebrate differences.

Attention

A related issue is the form of support given. When learners are temporarily stuck, it can be tempting to try to give them a push by telling them something that they could have worked out for themselves. By being aware of what you are attending to, and trying to work out, perhaps by asking questions, what they are attending to, it often emerges that there is a mismatch. Learners' thinking can be influenced by attracting or focusing attention, and by prompting relevant kinds of attention: discerning details and features, recognising relationships between features, perceiving relationships as properties so as to be able to reason on the basis of established and agreed properties. Often what is problematic is the form of learners' attention.

As a teacher it can be difficult to anticipate how learners will begin a task. Very often there will be surprises and new approaches. A further difficulty is to recognise whether an approach is legitimate, and to accept that when it is not, some good geometry and good problem-solving might still be learned.

Learners often answer the question 'why (does this property always hold)?' with the reply 'it just is' or 'it's obvious'. This sort of reply is what might be expected by the van Hiele (van Hiele and van Hiele-Geldof, 1986) approach and by thinking in terms of the structure of learners' attention. Learners may be focusing on relationships and properties but not yet isolating some properties as the basis for reasoning and other properties as susceptible to reasoning. Of course, learners may also sometimes say 'it's obvious' when they are trying to cover the fact that they are not sure and are only guessing.

Learning to reason involves both constructing your own chains of reasoning and also learning to listen to and critique other peoples' reasoning. Getting learners to work alone briefly, then in pairs, then in fours before having a plenary discussion can enrich the discussion at each level, and alert individuals to ways of thinking that had not yet occurred to them, as well as recognising that others are struggling in similar ways to themselves.

Task 9.5.1 Reflecting on Learning

Think about your engagement with this chapter.

What was entirely new to you? How did you deal with these new ideas? In what ways did you feel challenged? Did you feel that the tasks moved too fast, that you were being forced to rush on?

What was already very familiar to you? How did you deal with these ideas? Was your general feeling one of being relaxed in your comfort zone, or were you bored? Were the approaches used rather unfamiliar?

Did you have difficulty in classifying what was new and what was unfamiliar? Were you surprised about this?

Comment

All learners build on their prior experiences when learning. In addition, learning is never 'complete'. There is always more to learn about a subject in terms of seeing it from a different perspective, using it for a different application, and seeing similarities and connections with other topics and situations. Consequently, asking you what was entirely new and what was already very familiar is a little unfair: nothing is really entirely new, and nothing is completely familiar.

A mathematician is content to leave challenging questions as conjectures to be returned to at a later date. A good learner also is willing to leave some things they encounter as conjectures or as 'faint tastes' to which they might return in future.

Learners at school may not have ever thought about a one-way stretch in geometry, but they would have informal ideas linked to it; and they may confuse changing a square into a rhombus with a shear, but they would still have experience of both transformations.

Articulating

The tasks in this chapter often assumed that you would stop and develop an image (which might be mostly kinaesthetic–aural or mostly pictorial). You were often expected to draw your own diagram or even to arrange some physical manifestation. The point about finding something which is confidence-inspiring to manipulate is not to answer the specific question, but to enable you to get a sense of what might be going on, to recognise some relevant relationships and perceive appropriate properties. As this sense develops, it becomes easier and easier to express those relationships and properties in diagrams, words and symbols. As you become more articulate you begin to find yourself using those articulations as confidently manipulable entities for future sense-making.

Articulating usually suggests talking about something, and in a group learning situation this would be normal but might not leave any visible record. Working individually you may have written some notes, or you may just have thought 'Aha, I see', or even 'I sort of see so I shall move on'. Seeing ideas go by is rarely sufficient to enable you to reconstruct them later. Continued experience through multiple exposures to the ideas and ways of thinking can eventually lead to a growing sense of confidence and mastery. You can only say that you have truly learned something when you can explain it cogently to someone else. Hence the importance of going beyond 'seeing' relationships and the relevance of properties, to 'saying' what you see, expressing yourself eventually in diagrams and words and even symbols. However, it is important to actually say or write something as this seems to make the ideas more concrete. This articulation might be in a reflective diary where you not only jot down what you have 'got a sense of', but also how you might modify and use this task with your pupils.

Learning Styles

Learning styles are often discussed in schools and, while it is difficult for teachers to cater specifically for individual learning styles, you need to be aware of some aspects of them. You also need to decide whether, if a pupil has one well-developed learning style, it is your main responsibility to teach them all you can by catering for that style, or if you have responsibility to help them complement this style by extending their repertoire of styles.

Task 9.5.2 Thinking about Your Own Learning and Teaching

As you worked through the tasks in this chapter did you find the words or the diagrams most helpful? Did you list new words in tables or organise them in some form of schema? If you were teaching these topics would you use mainly words or diagrams and how might you modify your teaching to balance the two?

Comment
Some learners like to have a big picture then fill in the gaps, others like to start with the small details and build up the bigger picture. Which of these two approaches do you prefer? Think how a topic such as congruence transformations might be taught starting with the big picture, and how this would contrast with starting with the details.

Partly linked with the big picture versus small details approaches is a preference for analysis or synthesis. Do you prefer to explore a situation to find everything you can about it, or do you prefer to start with a range of apparently disconnected ideas and look for connections between them?

Again, using a task like congruence transformations, how might a lesson be structured emphasising analysis, and how might the same topics be presented if the emphasis is on synthesis?

When you meet a new idea are you satisfied to know what it is about, or do you also want to know why? For example, if you were told 'if three sides of one triangle are proportional to three sides of another, then the triangles are similar' you might first think, 'now what does that mean?' and then, perhaps, 'yes, I can remember that', or might you be wondering, 'why is that so?' Which reaction is more likely to occur with learners, and how might you promote asking 'why' rather than just 'what'?

When trying to think about an unfamiliar geometric idea do you find yourself making intuitive jumps or do you always work through the task one step at a time? Do you listen to your intuition? What do you see as the role of intuition in learning a subject like geometry? Do you encourage pupils in class to listen to their intuition? If you do make an intuitive jump, do you always check it before proceeding further?

These different perspectives on learning have been articulated as dichotomies, when really they are more like continua, with most learners being somewhere between the extremes at any one time. Being aware of the range of preferences that pupils have and being able to structure lessons so that different perspectives can be accommodated makes the teaching more effective and the learning richer. Knowing your own preferences is even more important as it is very easy when teaching to act out the implicit assumption that learners will learn as you do.

Reflection 9.5

Most people who read this book will either be familiar with Euclidean or transformation geometry, depending on when they went to school, but not both.

Which feels more comfortable for you at the end of this first chapter about transformation geometry? What have you left as a 'conjecture' or a 'taste' for the time being?

10 Language and Points of View

This is the third chapter focusing on points of view and language, this time in the context of geometrical transformations.

The first section is about paper-folding and tessellations.

The second section is about a transformation known as a *dual* which involves non-congruence-preserving transformations, and requires you to switch attention while working with tessellations.

The third section takes the idea of a dual into three dimensions and the fourth chapter looks at applications of the notion of dual. The chapter ends with a review of pedagogic issues encountered.

10.1 RECOGNISING ROTATIONS AND REFLECTIONS

This first section is about paper-folding and tessellations.

Different points of view can arise most unexpectedly in classrooms, but when recognised as such, can be exploited. Consider this anecdote from a teacher in an 11–16 school.

'The Vulture and the Mouse', by John Hancock

The all-ability Y7 classes were due to do some work on rotations and angles. The material in the textbook seemed accessible and the scheme of work indicated how it could be differentiated. I decided we would all start at the first question (figure 10.1a) in the 'Turning' section of the book. I planned to tackle the vulture and the mouse problem orally and take very little time over it.

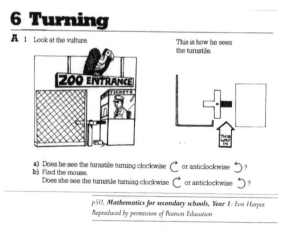

Figure 10.1a

The class studied the question: I directed their attention to the text, diagram and picture with the intention of ensuring everyone could understand the problem. I then asked for answers and the first two replies indicated differences of opinion about the correct direction. These differences of opinion were real: on asking for a show of hands it became apparent that there was a very large majority opinion in favour of one particular direction.

Students were becoming quite outspoken about the question and after a few minutes I invited them to form three groups: the clockwise and anti-clockwise factions faced one another across the room and two 'don't knows' came and stood by my side.

In the light of the size of these groupings I was forced to reconsider my own answer: the very few allies I had in the class were some of the weakest students in the year and most of the strongest mathematicians had confidently expressed an opinion which contradicted my own view! Students clearly expected me to arbitrate in this intellectual dispute but I remained silent as to the correct answer. I invited students to talk and listen to one another's answers. I was able to stand back and observed:

- Much on-task talking, listening and questioning. Opinions were expressed loudly, immediately, vehemently and freely. Students tried to persuade and convince others of the correctness of their answers. Disagreements about what IS clockwise were in evidence.
- Students attempting to simulate the problem. Some were standing on desks and others lying on the floor. There was a request to take down the clock. Some students built models. Students pointed at themselves and rotated their fingers and I was barely able to contain myself as they struggled to 'get behind' their hand in order to see the opposite view!
- Some changes of opinion, but not many. I remained, disconcertedly, a member of a very small minority!

This activity continued for the whole lesson. It did not appear to be a waste of time as the class were totally engrossed in the problem. The lesson seemed to have been a success despite the departure from the original plan.

Later that day I heard students talking in the corridor about the problem. Staff approached me to ask why their lessons are being punctuated by references to vultures and mice. The next day some students were wearing badges which proclaimed the direction they thought was correct – 'It's … '

I have shared the vulture and the mouse question with a group of teachers. Their initial response – a polite silence – was disconcerting. Had I missed something really obvious? Their reaction suggested they thought this to be a straightforward question and that students should have no difficulty in answering it correctly. I suspect they would have made similar judgements to my own about how to use it in the classroom.

I find this anecdote worrying. In a school climate where I am supposed to have detailed prior knowledge of student performance I cannot explain why this question proved so controversial to the students. I am also at a loss to know as to how I could have quickly and meaningfully sorted out the misconceptions.

I find this anecdote encouraging. It suggests that meaningful mathematics can be enjoyed by all students of all levels of ability engaging in a common task. I was pleased with the way students at my school engaged in the task.

I find this anecdote challenging. The question revealed deep disagreements between students and I am concerned about how easy it would have been to skip over what appears to be a real failure of understanding.

The anecdote could of course be worthless. My experience might be just an isolated case which will not be replicated in other schools. I could also have got the wrong direction!

John S. Hancock is head of mathematics at Settlebeck High School, Sedbergh, Cumbria.

(Reprinted (with permission) from *Mathematics Teaching*, 185, December 2003)

Task 10.1.1 Language and Points of View

John Hancock gives *an account of* what happened in his lesson. Make some brief notes which might *account for* what happened in the lesson.

How might you account for the silence mentioned by the author when he shared the question with a group of teachers? How could you break the silence if John had told this story to you?

Comment
In a mathematics classroom, what the teacher sees in his head as he introduces a topic can sometimes be very different to what, at least, some of the learners visualise in their heads. John was sufficiently aware to notice what was happening and had the skill to stay with the moment, teasing out different points of view until some common ground was established from which work on rotations could begin. But that was on another day that he does not tell

---

Approaching geometry from a transformational point of view was introduced into the curriculum in the 1970s in an attempt to make the geometry of Euclid accessible to a wider group of learners. Since affine transformations are conveniently studied using matrices, these were intended as a unifying theme. However, the work needed to use matrices to deal with transformations was too demanding and the project failed. Transformations subsequently disappeared from GCSE examinations.

One of the major issues in constructing a geometry curriculum is to find a logical basis from which reasoning can build. Euclid uses very simple axioms, but it takes a long time to get to interesting results. Transformations are difficult to reduce to a few axioms, but more closely match how people see the world: indeed, the word *theorem* comes from Greek meaning 'a way of seeing'. The fact that we operate in time and space, and that, while time can be described algebraically, space is described geometrically, means that geometry and algebra are complementary, even opposite sides of the same coin.

The building blocks of geometry are dynamic images. Indeed, geometry has been described as the study of the dynamics of the mind, for example by Caleb Gattegno (1987). The electronic world we inhabit is increasingly driven by images, both static and dynamic, and the study of geometry can help learners to develop their powers so as to be able to make sense and use of these images.

Every diagram in a geometry text is a particular instance of a general case. The work needed to make sense of it is to become aware of what is permitted to change and what relationships are necessarily invariant. The diagram provides a stable background to enable visualisation or a sense-of generality, of properties which must hold, or which cannot hold, and of chains of reasoning using those properties. This is true whether thinking in Euclidean or in transformational terms.

Seeing what is drawn or saying what is seen (construing) and seeing what is said (imagining) are two of the geometric powers that young children work on long before they meet formal geometry in school. If you have any way of observing pre-school age children it is worth watching a child of 3 or 4 working directly on construing. For example,

> We had a family outing to Talkin Tarn a couple of months ago. We parked the car and walked to the tarn edge by the boathouse and turned right. Soon Anna (age 3 years 2 months), Anna's mum, and I were on our own behind everyone else. Stile climbing and wellington waterproof testing to the limit were the order of the day. We moved slowly on and in and out of the water's edge. Anna on top of a stile looking across the tarn said 'There's a boathouse'. Anna doesn't expect replies. You don't often get space to reply. We walked on. More water play. Wellies wet inside and out by now. Two stiles later an excited 'There's another boathouse.' The chatter of saying what is seen, testing her construing of the world remains in my memory alongside the sensitive interplay with her mother's language, when she at last managed to get a word in, and remarked as we got back to the car 'Haven't we seen the boathouse lots of times on our walk.'

This, non-judgmental, adult interjection that her many boathouses might only be one, seen lots of times, is a lovely example of the complexity of imagining (seeing what is said) and the interplay with saying what is seen (construing) when you are 2 or 3 years old.

As you work on the next tasks, which involve paper-folding and cutting, pause occasionally and ask yourself whether you make connections between the task and any of the plane transformations: reflection, rotation or translation.

Task 10.1.2 Paper-Folding and Cutting

Cut several pieces of A4 paper in half lengthwise Take one piece. Fold it in half and half again with the two folds parallel and parallel to the short side. Draw a man on the quarter piece in such a way that when you cut it out you will be able to produce a string of four men holding hands.

Comment
This task was deliberately posed in words, not pictures. Presenting it this way ensures that it is an imagining task. If you struggled with the words then the pictures in the next task might be a help in getting you started.

Task 10.1.3 How is it Done?

Take a new strip. Is it now possible to fold, draw and cut so that you can unfold to make a string boy–girl–boy–girl holding hands as shown in the first picture in Figure 10.1b?

Figure 10.1b

What other strings could be made from folding and cutting a strip of paper, and how were the other ones in the figure done?

> *Comment*
> Recognising and acknowledging being stuck is an important state, as it makes it possible to learn. When you cannot see how to do something, you have to ask whether it is possible. In this case, three of the four are indeed possible. Useful questions to ask include 'what do I know?' and 'what do I want?' As was said previously, it rarely takes long to answer these questions. What you are seeking is some sort of relationship or route between them.

It is tempting, when you are really stuck, to dismiss the task as inappropriate for your own learners. But it is astonishing what learners can achieve when challenged. Sometimes a group will come up with an approach that you did not think of on your own.

In this case, since you know that the idea is to fold and then cut, you might as well search for alternative ways to fold. You can take the point of view that the task is about folding and cutting. You can also take the point of view that the task is about symmetry or about reflection. Each of these viewpoints offers different insights, different ways of working on the problem.

If the task is about folding and cutting then there are questions about whether to concertina the paper or fold over and over. If it is about symmetry, do you fold an extra time and draw half a person? How do you place that half person on the folded piece? But this starts to impinge on folding questions; for four figures do you need to fold to get eight thicknesses of paper and why did those half people appear at the end of the line on the first try? Can this be sorted out by using some different concertina folding?

If it is about reflection, where are the mirrors? Are there mirrors down the middle of each person and/or between each person? What happens after two reflections in two different mirrors?

Most importantly, this task is an opportunity to manipulate, to get a sense of, all of the above. Only after working at the task will some of the writing start to make sense.

Task 10.1.4 More Folding and Cutting

The eight some reel shown in Figure 10.1c was made by folding and cutting a square sheet of paper, although it could equally well have been made using a circular piece of paper.

What dimensions of possible variation are available in this task?

Comment
You could vary the number of occurrences of each type, and the number of types.

Many mathematics texts offer 'ink devils' as an activity but in an apparently random way as something to do at the end of

Figure 10.1c

term or after they all come in late from PE. They can be found in textbooks for Year 3 right through to Year 9. But 'ink devils' offer an opportunity to work on detecting and recognising reflections. In the boy–girl–boy–girl task every fold is a mirror. Does this 'point of view', this manner of attending, need to be made explicit, either to you or to a class? And if the answer is 'yes', then when does it need to become explicit?

Asking learners to take a different point of view before they are comfortable with what they need to solve the problem can be counter productive. Not drawing the idea of reflection into their consciousness at some stage could miss an opportunity. The mirrors in Tasks

> 10.1.2 and 10.1.3 are all parallel. The mirrors in Task 10.1.4 are at an angle. Comparing the tasks: reflection in two mirrors gives a translation when the mirrors are parallel (10.1.2 and 10.1.3) and a rotation when the mirrors are at an angle (10. 1.4).

'The mathematics classrooms had no displays on the wall to indicate that this was a place where mathematics was being taught' (Ofsted report). A response to requirements of this sort can be to produce a quick display of ink devils. The learners have had fun in constructing them, but were they involved in mathematical thinking?

To appreciate the relationship between translations and rotations, it is necessary to become aware of the difference in effect between having two mirrors at an angle and two mirrors parallel. You learn new technical terms most effectively by having access to a range of examples and a sense of what can vary and still something remains an example.

Many more questions can be asked, such as:

- 'If an ink devil is made with two folds, can it have more than two axes of symmetry? Could it have three?'
- 'Can a five-headed gorgon be made by folding?'

The next tasks stay with the same content, but vary the context. You are asked to use dynamic geometry software to work with mirrors and reflection. When using dynamic geometry software, it is easier to draw geometric figures than boys and girls, so a square, an octagon and a hexagon have been chosen as the starting shapes. What is not changed is that all these polygons have an axis of symmetry parallel to the mirror line.

Task 10.1.5 Tessellating Shapes

Use the interactive file '**10a Simple Quad Tiling**' to explore the construction of the tiling indicated in Figure 10.1d, and the interactive file '**10b HexSq Tiling**' to explore the tiling shown in Figure 10.1e.

Starting from one tile or one pair of tiles, what transformations do you need to apply to generate the next (adjacent) tile, and the next, and the next, and so on?

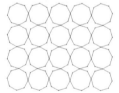

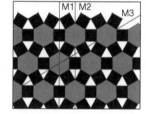

Comment　　　　　　　　　　*Figure 10.1d*　　　　　　　*Figure 10.1e*

You can think in terms of always mapping the starting tile, or of acting on one of the intermediate tiles. Since you are only allowed to use mirror reflection starting with just a few mirrors, you may want to reflect a mirror line as well to use later.

The previous task involves explored tessellations based on tiling. Tessellations of squares and rectangles are very evident in the world around.

Notice the differences between working with paper-folding and scissors, and dynamic geometry software for promoting thinking about reflections and rotations and translations. Different people find different media easier to work with and to learn from.

The hexagon and square tiling highlight a problem in working with figures with an axis of symmetry parallel to the mirror. It can be difficult to appreciate the difference between a reflection, a translation or a rotation. The object on the left of Figure 10.1f demonstrates the way that symmetry can disguise a transformation. The square from the hexagon–square–triangle tiling has been replaced by a scalene triangle.

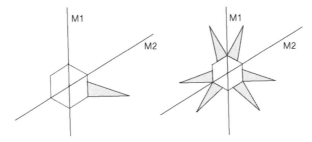

Figure 10.1f

Reflecting in the two mirrors shown produces something which, with a bit of imagination, might be seen as a rabbit with ears, arms and a pair of splayed feet. The new figure now has symmetries that were destroyed when the single triangle was added. The new symmetries are due to the fact that the composition of reflections in the two mirrors is a rotation through 120°. However, further reflections in either mirror adds nothing new (the two mirror reflections, the rotation and the identity do-nothing transformation form a closed collection of transformations).

You might like to open and explore the interactive file '10c Kaleidoscope' at this point.

Not all patterns are tessellations, or put another way, tilings can emerge from overlaps between tiles that do not themselves tessellate.

Task 10.1.6 Patterns Made with Regular Pentagons

For each of the patterns shown in Figure 10.1g, find some feature which distinguishes it from the other two, that is, something which the other two share but it does not.

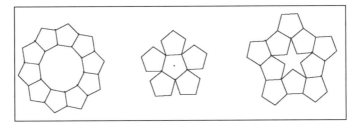

Figure 10.1g

Comment
It is often easier to find differences when discussing with others what you see, that is, the features you are attending to, and the features that others are discerning. Alternatively, say to yourself, or even make a note of, various features and details that you can discern.

Task 10.1.7 Tiling with a Regular Pentagon

How many different ways can you find to tile the plane with a regular pentagon and one other shape?

Comment
Although no definitive count will be given, a variety of possibilities will emerge in what follows. As with many tasks, a night's sleep often produces a useful conjecture or starting point the next morning. Insight cannot be forced, and learning is a process of maturation which takes place over time, not in an instant during a lesson!

Figure 10.1g shows why regular pentagons do not tessellate. But together with regular 'gaps' thought of as tiles, they can be seen to tile the plane, as in Figure 10.1h.

Figure 10.1h

Figure 10.1i was constructed by taking the middle figure in Task 10.1.6 and building symmetrically around a central starting point. As the pattern builds it seems to stumble upon both the other arrangements.

Figure 10.1i

Constructing central tessellations can be a very powerful activity. It appeals to order-making and symmetry, recognising areas deep in the psyche. It leads to a feeling that you have control of what you are creating. However, with a non-tessellating polygon (pentagon, heptagon, nonagon), there is the additional excitement of whether the pattern will continue for ever to infinity. Figure 10.1j shows a photograph of people using pentagonal mats to form such a central tessellation.

As designs build up physically there is a palpable sense of experiencing the symmetries and noticing when they are broken by mistake. Using dynamic geometry software often leads to a considerable amount of 'undoing' in order to retain the symmetries

Figure 10.1j

Some pentagons do tessellate (they have all been catalogued and there are hundreds!). The one shown in Figure 10.1k is often referred to as the Cairo tessellation because it appears in a mosque there.

As you look at the tiling, you may become aware of switching your attention between pentagons and hexagons. The hexagons make it easier to see why the tiling can be extended indefinitely. There are tilings of the plane which are completely irregular. One group can be found on the web by searching for 'Penrose tiling'.

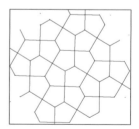

Figure 10.1k

Nesting, Stacking and Alternating

Chairs and tables are often constructed so that they will stack easily. The same idea is useful when trying to see whether a shape will tessellate: you look for a larger grouping for which the transformation is easier to 'see'. A similar approach is useful when looking for different ways to tessellate. Take, for example, a T-shape as shown in Figure 10.1l.

Figure 10.1l

This T can be nested, stacked and alternated, as shown in Figure 10.1m.

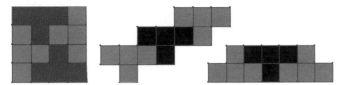

Figure 10.1m

Figure 10.1n is a tiling using pentagons and one other shape. By stressing a certain sub-figure and treating it as a tile, it becomes clear that the tiling continues indefinitely. Other tiles can be found by replacing one of the component pentagons with a different adjacent pentagon.

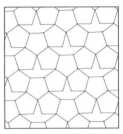

In this section you have worked with translation, reflection and rotation; you have seen that each of these preserves length, area and angle. In summary, they preserve congruence. There are other transformations of the plane that you have met that do not preserve congruence. Enlargement preserves similarity but not congruence. A shear preserves area but not congruence.

Figure 10.1n

Reflection 10.1

Did you find yourself wishing you had worked on this section before the previous chapter? Did you find that the work in this section helped to make sense of and connect up ideas from the previous chapter? What does that suggest about planning activities for learners?

10.2 DUALITY: A NON-CONGRUENCE-PRESERVING TRANSFORMATION

This section moves from congruence-preserving transformation to non-congruence-preserving transformations to demonstrate another way of looking at tessellations. You will work with *duality* as a transformation in two dimensions and then, in the next section, extend the idea of duality to three dimensions. Work through the tasks but use the tasks to ask yourself about the mathematics behind duality. What is being preserved in a transformation like this? What is not preserved? What can I know by looking at a tessellation and its dual as a pair?

The three patterns in Figure 10.2a were made by allowing overlaps as a regular polygon is reflected in each of its edges.

Heptagons   Octagons   Nonagons

Figure 10.2a

Considering heptagons more fully, the process of reflecting can be repeated on every edge, or just on the outward-facing pair of edges, to produce a sequence of layers. It takes a moment or two to see why the overlaps, treated as a tile, keep reappearing in each layer without distortion.

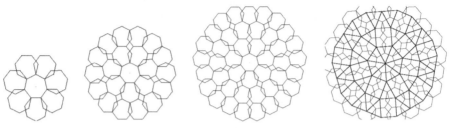

Figure 10.2b

A second pattern emerges by joining the centres of each heptagon to the centres of the adjacent heptagons, known as the *dual* tiling, as shown in the diagram on the right of Figure 10.2b

Task 10.2.1 Heptagons and Duals

Did you think to check whether the angles and the sizes remain the same as the pattern spreads outwards? What do you need to do to convince yourself that the dual tiling really only uses two tiles (that is, that the sizes remain the same for ever as the pattern extends outwards)?

Write down instructions for someone to sketch the heptagon 'tiling' and for someone to sketch the dual tiling (without using the heptagons).

In the dual, there are seven triangles in the first layer, and seven triangles and seven rectangles in the second layer. How many of each size tile will be in the *n*th layer?

Comment
There is at least one way of seeing the dual tiling as a tessellation with hexagonal tiles made up from triangles and rectangles. Notice how an algebra-generalising question can emerge from contemplating a geometrical pattern.

When you start with a tiling (*not* an overlapping as with the heptagons) and construct the dual, you get another tiling and, rather elegantly, if in turn you construct the dual of that tiling, you get back to the first. Hence the name *dual*.

Task 10.2.2 Generalising the Cairo Dual

Using the Cairo tiling, draw its dual using Figure 10.2c. Convince yourself, and preferably someone else, that there are relationships between the numbers of edges adjacent to a vertex in a tiling and its dual, and the number of edges of the tiles.

Comment
You can use information about the dual in order to reason about the symmetries and angles of the pentagons in the original tiling.

Figure 10.2c

Figure 10.2d shows yet another tiling using regular pentagons and one other shape. This one might be unexpected because the pentagons do not meet along a full common edge. It makes use of glide reflections to get from one line of pentagons to an adjacent line. Its dual tiling is made up entirely of triangles.

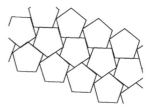

It is worth exploring whether this is the case for other tessellations of a pentagon and one other shape.

Figure 10.2d

10.3 DUALS IN THREE DIMENSIONS

This section is based on the use of materials which support construction of surfaces in three dimensions, such as ATM mats (see References) and Copydex, or plastic construction kits.

In preparation for the next task, you should, if at all possible, make a cube using six squares. It is worth spending a little time working initially with a familiar shape so that later you are not frustrated by the technical side of model-making and are able to concentrate on making what you have imagined.

Task 10.3.1 Truncating a Cube

Imagine a cube. Imagine that it is made from something like hard cheese that you can cut, in your head, with a sharp knife. Imagine it sitting on the table in front of you. Look at the top corner nearest to you. Cut off this corner making an equal cut on each of the top two edges and on the edge down the side.

What shape do you imagine the cut surface on your cube will be? What shape have you cut off? In your imagination make exactly the same cut on each of the other three top corners then turn your cube over and cut off the four corners in the same fashion. Imagine the object when the lengths cut off at each vertex are equal to the length remaining on the edge of the cube.

If possible, make yourself one of these objects.

Comment

This task has been recorded as an **Imagining** on the CD-ROM.

Were you able to adjust the size of the piece cut off in your mind by changing the amount of each edge cut off? Notice that in your mind you can make variations, such as arranging that the octagon remaining from each face is regular (equal sides and equal angles) by cutting off just enough of the cube edge. The polyhedron is known as a *truncated cube*. Making one physically may require calculations!

Task 10.3.2a Constructing a Cube Octahedron

Now imagine cutting off bigger and bigger slices from each corner until the slices meet in the middle of each edge. What shape is left on the face of the cube? You can make this polyhedron from squares and equilateral triangles. You might want to draw a net as well. In either case it is helpful to predict the number of squares and the number of triangles needed.

Comment
Notice how you can do the counting in your head without actually 'seeing' or mentally touching each face!

Compare your results with the illustrations in Figure 10.3a

Figure 10.3a

To see why this last solid is called a *cube octahedron*, it helps to make two further models.

Task 10.3.2b Constructing an Octahedron

Take eight equilateral triangles and use them all to make a solid that has exactly four triangles meeting at each vertex. Your model will have eight faces, and is known as an *octahedron*.

Comment
Did you start gluing, or snapping pieces of Polydron together, or did you pause first and imagine the polyhedron? Does thinking in terms of two square-based pyramids fitted base to base help?

The Platonic solids are the five regular polyhedra: in each Platonic solid, the faces are all the same regular polygon – equal sides, equal angles – and the same number of faces meet at each vertex. These solids were known to Plato around 500 BCE, and played a significant role both in Greek philosophy as metaphors, and in Greek geometry because it was possible to prove that there were only five of them.

Tetrahedron: 4 equilateral triangular faces, 4 vertices, 6 edges, 3 faces at each vertex
Cube: 6 square faces, 8 vertices, 12 edges, 3 faces at each vertex
Octahedron: 8 equilateral triangular faces, 6 vertices, 12 edges, 4 faces at each vertex
Icosohedron: 20 equilateral triangular faces, 12 vertices, 30 edges, 5 faces at each vertex
Dodecahedron: 12 regular pentagonal faces, 20 vertices, 30 edges, 3 faces at each vertex

All five can be found carved in granite by neolithic people living in Scotland.

Task 10.3.2c Truncating an Octahedron

Imagine an octahedron. You might like to consider, in your head, an ancient pyramid sitting so close to the River Nile with the Nile so still that you can see the reflection of the pyramid in the water. This gives four triangular faces above ground and four virtual triangular faces below ground.

Now cut off the top of the pyramid to produce a square face. In your head, what has happened to the bottom of the reflection in the Nile? Now cut off one of the corners on the ground in the same way, watching what happens to the reflection. Go round all the other base points of the pyramid doing the same thing.

Comment
Some people have to work to see that all the vertices look the same: the vertices at the top and the bottom often seem to be more pointed that the vertices around the 'middle'. But making one physically and turning it around corrects this impression.

Truncation as Transformation

Think of the five polyhedra as forming a sequence:

  Cube   Truncated cube   Cube octahedron   Truncated octahedron   Octahedron

If you imagine the truncation of the cube as an action on the cube which starts at the vertices and gradually moves along the edges until the shape becomes a cube octahedron, and then continuing (the cutting planes move through each other) to end up at the octahedron, you see that to each face of the cube there is a vertex of the octahedron and vice versa. Both have 12 edges. Each is said to be *the dual* of the other.

Task 10.3.3 Truncating a Tetrahedron

What do you think will happen if you truncate a tetrahedron? Now actually imagine it happening.

Figure 10.3b

Comment
At one position you have hexagonal and triangular faces, then suddenly you get all triangular faces (Figure 10.3b). Pushing further returns to a tetrahedron again: the tetrahedron is self-dual.
  Another way of thinking is to see the tetrahedron with tetrahedral caps cut off leaving an octahedron inside. Or put another way, start with an octahedron, select four faces, no two sharing an edge, and put tetrahedral caps on each of those four faces. A principal pedagogic purpose of working with polyhedra is to develop imagery in three dimensions, and to experience thinking about the same object in very different ways.

If you would like to take these ideas further you could consider what happens when you truncate a cube-octahedron: how many of each type of face there would be, how many vertices of each type (number of incident edges). See for example Cundy and Rollett (1981, p. 106).

Task 10.3.4 Tetrahedral–Octahedral Links

Imagine a tetrahedron contained in a sphere, seen as a large transparent balloon that just touches all four vertices of the tetrahedron that is trapped inside. Join each vertex of the tetrahedron to the mid point of the opposite triangular face and extend this line out until it touches the sphere. This will give you four new points on the sphere. Imagine them as the vertices of a new tetrahedron. Describe to yourself how the two tetrahedra intersect.

Once you have finished the task, make a model and review your imaginings.

Comment
The process of projecting centroids of faces onto a sphere works quite generally for convex polyhedra, and produces the dual polyhedron.

An alternative viewpoint is achieved by starting with an octahedron (which is the common intersection of the two tetrahedra) and adding to each face a tetrahedral *cap*. The result is known as a *stellated octahedron* or *stella octangular*. Looking at a physical model reveals two interlocking tetrahedra as in the task.

The stellated octahedron has a number of properties which are best verified by looking at a model:

- the solid common to the two tetrahedra is an octahedron
- the vertices of the stella octangular are the vertices of a cube
- the edges of the stella octangular are diagonals of the faces of the cube and intersect at the vertices of the octahedron
- the volume of the cube is six times the volume of the octahedron
- the volume of the stella octangular is twice the volume of the octahedron.

Relative volumes can be calculated by noting that the large tetrahedra are scalings of the small ones by a factor of 2 on the edges, and so by 8 on the volumes. But a large tetrahedron is made up of four small tetrahedra and one octahedron. So the octahedron is four times the volume of the tetrahedron. This hardly seems believable. It is a good example of how difficult it is to estimate volume (or area) by eye. It follows that the volume of the stella octangular is twice the volume of the octahedron.

It proves to be powerful mathematically to shift attention to studying processes as a way of studying properties of the objects. It is powerful pedagogically to immerse learners in experiencing such shifts, because the residue of their mathematical experience will not be particular facts about octahedral or other objects but, rather, a growing awareness that there are 'facts' and that these can be justified by reasoning on the basis of known and agreed properties. It is experience of ways of thinking which is valuable.

Reflection 10.3

What do you need to do in order to shift from thinking about individual polyhedra to thinking about the processes of constructing the dual and stellating?

10.4 APPLICATIONS

The geometrical content of this chapter has included transformations in two and three dimensions. You used transformations to look at tessellation and duality to make links between two- and three-dimension transformations. Behind much of the work was the idea of an infinite Euclidean plane and infinite three-dimensional space. Where are the applications of these particular geometric ideas?

There are two related areas to look for answers to that question. One is within mathematics itself, the other is within the material world of designing, making and doing.

Material World Occurrences

Wallpaper with a continuous repeating pattern is a comparatively modern phenomenon. The invention of large rollers and a continuous printing press opened up a whole new world of possibilities. Designers like William Morris, were quick to combine technology with mathematics. The mathematics told them that rectangles, equilateral triangles or hexagons tessellated the plane but also that any quadrilateral or any triangle also tessellated the plane. Your work in section 10.2 of this chapter allows you to make statements about the conditions for pentagons to tessellate the plane. Designers use this information in order to place flowers, birds and animals to ensure a visually pleasing result. Your work on stacking, alternating and nesting lies at the heart of wallpaper design. Stacking will always work. Alternating and nesting motives that include animals or people can lead to disturbing images with figures turned upside down or placed in unnatural positions. These are the results of transformations, so it improves a designer's flexibility and creativity to be aware of the mathematical relationships and properties.

The artist M.C. Escher is particularly well known for his explorations of geometrical ideas. He works within the world of mathematics making the mathematical ideas of infinity, continuity and transformation available to a wider public. His work includes explorations of Möbius bands, continuous impossible staircases, curved space, general techniques for tiling both Euclidean planes and non-Euclidean spaces (mentioned in Chapter 8).

Telephone cable, string, climbing ropes, lines for tying up boats, and steel cables for suspension bridges and cable cars are all made from twisted strands. Weaving strands together ensures elasticity as the load changes, as long as the strands are in contact with each other in a symmetrical fashion. The strands have to be arranged so that they are in maximal contact with each other, which leads to strands made up of strands made up of strands. Each stranding consists of three fibres (each is in contact with the others, which is not the case for four or five) or seven (one central and six surrounding) or nineteen (another layer in a hexagonal packing), and so on.

The study of how molecules and crystals fit together, and why they take the shapes they do is geometric at heart, building on and exploiting transformations to form tilings. Indeed, a good deal of geometry arose from chemists' and crystallographers' observations. Fullerene, named after Buckminster Fuller who extolled properties of the cube octahedron, was discovered as a molecule only in the late twentieth century.

The mathematical notion of a dual tiling or dual polyhedron is an example of a powerful way of thinking which supports a shift of attention from a particular object or pattern to becoming aware of relationships, enabling properties to be perceived which might then apply to other features or to other situations. Denoting each tile by a single point, and joining points if they have some specified relationship (in the case of dual, of being adjacent) extracts the relationship from the context and enables mathematicians to pose and resolve problems which are only to do with the fact of a relationship rather than the specifics of that relationship. The study of the configurations which result is known as *graph theory*. A good example of this is the famous design of the map of the London underground. It stresses relationships of connection, rather than geographical spatial relationships.

Mathematical Occurrences

Shifting attention from transforming objects to the study of transformations as objects marks a major advance in mathematical thinking associated with the twentieth century (but, of course, with roots well into the past). The topics in this chapter (duality in two and three dimensions, stellating and truncating polyhedra) have offered no more than glimpses of fruitful explorations involving the Platonic solids, other polyhedra, as well as objects in higher dimensions.

Some people feel that their intuitive appreciation is somehow diminished when they look at the world mathematically, but many people find that it enriches their appreciation because it brings up knowledge in unexpected places, forming a richer network of associations and connections. One of the complaints made about mathematics lessons is that learners are unable to see what connections there might be with other situations, whether mathematical or non-mathematical. Teachers who themselves were aware of connections between topics, and with non-mathematical occurrences, were found to be more pedagogically effective (Askew et al., 1997).

Reflection 10.4

Set yourself to look out for the occurrence of geometrical patterns. How does your appreciation alter when you ask yourself about which transformations could be used to generate the pattern, or what the dual might be?

10.5 PEDAGOGIC PERSPECTIVES

In thinking geometrically there will always be a complicated web of interconnected concepts, skills and facts that any one person brings to bear on a particular problem. For this reason it is both undesirable and unobtainable to strip the teaching of geometry down into a collection of isolated facts and skills to be known or practised.

In many of the tasks in this book, you may have experienced being stuck, perhaps even wanting to give up and 'look up an answer'; turning to some familiar object (physical, diagrammatic, symbolic) in order to manipulate it to get a sense of relationships; finding yourself prompted to look at something from a completely different viewpoint (truncating and stellating); making conjectures and trying to find reasons to justify those conjectures. You may have noticed times when what started as an action performed (a 'doing') was then reversed, into an 'undoing' or when an action became itself the subject of analysis (forming a dual, stellating, …). You may have noticed yourself drawing on knowledge of symmetry; shifting from thinking of a transformation as acting on an object (for example, a tile) and thinking of a transformation as acting on the whole space so that its invariances can be made use of. You may have felt some desire to explore what happens if you repeatedly stellate a polyhedron. You may have suddenly been aware of a connection, such as that the cross diagram of squares on the sides of a triangle is a form of stellating in two dimensions.

The important thing to emerge from all this is that a task worth doing affords opportunities for multiple outcomes. Each task has its outer aspect (what you are asked to do) and a range of inner aspects: mathematical themes you might encounter; personal propensities that might be available to work on; powers that might be called upon or might even be developed during the work; connections with other topics and with other non-mathematical situations which might arise; opportunities to rehearse and perfect previously met techniques; and experience of different ways of viewing something familiar.

Transformations provide a rich context in which to become aware of the multiple potentialities of tasks and situations. They provide a way of thinking about objects (which transformations contribute to appreciating the structure of the object, the way in which it can be generated or built up). They also provide a means to extract essential structure from an otherwise complex situation in order to resolve some problem or exploit some relationship as a property applicable in other situations. For example, seeing rotations as rotations, and also as the composition of reflections, helps in making sense of a tiling pattern. Similar experiences occur in other contexts, including arithmetic and algebra.

What makes a task pedagogically effective lies not so much in the task itself, but in the ways of working which have been developed. Pausing after work on a challenging task, and thinking back over where you got stuck and how you got unstuck again, what dimensions of possible variation are available, what connections there might be with other experiences, is much more valuable and efficient than diving into yet another task.

Chapter 13 proposes a structure for thinking about mathematical topics. One of the three component strands, corresponding to the structure of human psyche, is connection with and harnessing of emotions. Each mathematical topic arose from people trying to resolve problems. Once developed as a topic with techniques and ways of perceiving and thinking, there are many different situations in which the topic arises or in which the techniques prove useful. Learners' emotional involvement is enhanced when they experience themselves using and developing their own powers of sense-making, and realising the range of problematic situations from which a topic derives and in which a topic has been found useful.

What Makes an Example Exemplary?

In order to communicate a concept, it is useful and usually necessary, to offer an example and to engage the learner in activity. But, in order to appreciate what is being exemplified, it is necessary to have a sense of the generality of which the example is a particular case; in order to appreciate the activity, it is necessary to become aware of what generalities are being experienced in the particular example. In order to integrate and connect ideas, it is necessary for those ideas to be brought to attention, but if someone else brings them to your attention, you may not experience the power of 'seeing a connection'. This 'pedagogic paradox' points to the value of developing a perspective in which teaching does not guarantee, does not even produce learning. Rather, learning comes about as people make use of their own powers of sense-making to form connections, to generalise and to particularise, to build up a rich collection of familiar objects as examples which they can use to manipulate in the future when they want to resolve a problem, test out a conjecture or look at 'what is going on'.

For example, the paper-folding tasks at the beginning of the chapter depend on shifts between translation itself and translation as the composition of two reflections. If you are told how to do the cut-outs, you are unlikely to experience the delight which arises from using your own powers to 'really see' the connection. The particular example becomes, for you, exemplary. Activity you have undertaken in which you have made a connection or achieved an insight is as much to do with how you approached the task as the task itself.

Reflection 10.5

Create a concept map of your current understanding of duals and their relationships to other aspects of transformation geometry.

Explore the term 'dual' on the web and then note what changes you might make to your concept map.

11 Reasoning with Transformation Geometry

The first section in this chapter is a reprise of some of the ideas in Chapter 7, with a stronger focus on transformation geometry.

The chapter starts with suggesting alternative methods for proofs of Pythagoras' theorem. Three-dimensional shapes are introduced and relationships are sought using transformation methods. The triangle 'centres' are explored; the centroid is used to prove that the centre of gravity of a triangle lies at the point of trisection of medians and the orthocentre is used to find the altitudes intersect at the same point.

The third section offers some examples of undoing tasks in geometry, associated with centres, starting with the reversal problem between orthocentres and incentres.

The fourth section moves between Euclidean, measurement and transformation-based perspectives. There is further work on Vecten's diagram.

The fifth section is a review of pedagogic notions in the chapter

11.1 REASONING USING INVARIANCE

One way to demonstrate Pythagoras' theorem is to make copies of the squares on the sides out of coloured paper and to use the coloured paper squares to cover the square on the hypotenuse, cutting as required. This is not yet a proof but may form the basis of one if it is possible to generalise the description of how the cutting is to be done.

In Chapter 7 there were four proofs of Pythagoras' theorem, all based on a Euclidean approach, reasoning about incidence and results about congruence, area and so on. The proof offered by Euclid, was based upon the diagram in Figure 11.1a, though it must be said that diagrams were only introduced when Euclid was translated into English and other languages.

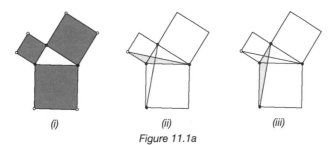

| (i) | (ii) | (iii) |

Figure 11.1a

At one point in the reasoning, it is claimed that the two overlapping triangles in part (ii) of Figure 11.1a are congruent. How might Euclid have found a route to proof with a step like that in the middle? To find the two triangles requires inserting some extra lines, discerning something which might, and in this case, does reveal a useful relationship.

A transformational approach might proceed as follows:

What is wanted? The area of the square on the hypotenuse is the sum of the areas of the squares on the other two sides.

What is known? The triangle is right-angled; areas of triangles are preserved by shears (in Euclidean geometry, triangles on the same base between parallel lines are equal in area).

Cut the first square into two and focus on the lower triangle (whose area is half the area of the square: how might you justify this?). Being aware of the right-angle brings to mind the fact that an edge of the square is an extension of an edge of the original triangle (again, how might you justify this?). Shear the triangle along that edge (See Figure 11.1b.)

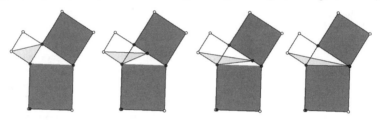

Figure 11.1b

Rotate the triangle about one corner, preserving its shape, until it is in the position shown in the third diagram in Figure 11.1c. This is intended to be experienced as a movie.

How could you justify the assertion that the triangle ends up with its edges matching the corresponding squares?

Figure 11.1c

Next draw in a perpendicular to the base of the original triangle (Figure 11.1d) and shear the triangle down this line (parallel to the side of the square on the hypotonuse).

Figure 11.1d

The Euclidean reasoning uses congruence of triangles rather than shearing.

Now the pieces can be assembled: the final triangle is half the rectangle that encloses it, so the starting square is equal in area to the rectangle. The same reasoning can be applied to the other square to reveal that (area of) the square on the hypotenuse is the sum of the squares on the other two sides.

You may have found that there were far too many words, or too few, between the pictures. Proof in transformation geometry is a case of manipulating visual images until you have a sense of the result. When you are able to articulate that result you are in a position to present a proof. Task 11.1a provides an opportunity to review the reasoning, both by completing the similar reasoning on the other square and by relating the verbal presentation with static pictures to a dynamic presentation.

Task 11.1.1a Rehearsing the Reasoning

Use the diagrams in Figure 11.1d to rehearse the reasoning on the second square.

At your first opportunity, go to the website Pythagoras (webref) and watch the Java applet you find there. What do you have to do in order to 'see' the applet as providing a convincing proof?

Comment

You may have found that the words used to state the reasoning in this text, while providing direction as to what to discern, relate and what properties to perceive, also get in the way of the basic perception of the dynamic images. The dynamic but wordless images provide the insight and experience to appreciate the reasoning. However, even if you have the facility to produce Java applets, there is more work to be done. In order to have a fully convincing proof, you need to be aware of the generality, the dimensions of possible variation implicit in the animation, and you need to check that in all cases the transformations do actually work (why do the angles work out, why do the lengths work out, …). It is possible to convince yourself sometimes that something works, when in fact it only works in special cases.

Task 11.1.1b Rehearsing the Reasoning Again

You saw Figure 11.1e in Chapter 7. What modifications are required to the reasoning you met there to prove the more general result that in Figure 11.1e, the areas of rectangles with the same shade of grey are equal?

Figure 11.1e

Comment
Did you find yourself 'seeing' the transformations as if imposed on the diagram? Each of the pairs of shaded rectangles can be treated the same way as the square and rectangle pairs in the previous reasoning.

Did you think to imagine the original triangle varying, so as to be aware of the dimensions of possible variation and the range of permissible change? A dynamic geometry diagram enables you to drag and vary, but this is only an intermediate stage to being able to alter a diagram mentally as you look at a single static version.

Transformations do not always provide the simplest approach to developing a chain of reasoning: seeing why the four triangles are all equal in area can be done with measurement or as transformation, by cutting the central triangle along each median in turn, then reassembling the two pieces to form the corresponding flanking triangle.

The reasoning in the last task uses shears, but simply saying 'shears' does not reveal anything. The property of shears used as the basis of reasoning was that shears preserve area of triangles (and hence quadrilaterals and other regions made up of triangles, including, using sophisticated arguments, areas of shapes with curved boundaries). This could be taken as the axiomatic or defining property of shears: shears preserve area.

Triangles are fundamental building blocks. Quadrilaterals are much more floppy. Euclidean geometry forces you to break down quadrilaterals into triangles and use congruence arguments. Transformation geometry can arise from experiences of handling elastic bands on geo-boards, from dragging images in dynamic geometry software, and from gazing at still images and imposing dynamic movement on discerned sub-figures.

The ancient Greeks did not have easily available pages and pages of paper. Papyrus was expensive and writing and drawing took time. There is a case to be made that one of the drivers of the way Greek geometry developed was the need to be able to work 'in the head'. This is why there have been so many tasks of the form 'imagine'. Almost any geometry task can be initiated with 'imagine a … ', each time challenging learners to develop their power to imagine a little bit further, and making use of static images as the background for further imagery.

Pythagoras' theorem has appeared and reappeared throughout this book. Which, of the earlier proofs, can you now recall and reproduce as a mental image that you could communicate to another by words alone? You may find that what comes to mind is something with which you are most familiar, and which uses your powers most fully.

Staying with the idea that triangles are a basic building block the next task increases the complexity by assembling isosceles acute-angled triangles into a polyhedron. You may want to build a physical model, using ATM mats: you can produce four identical isosceles triangles either by cutting up four pentagon mats or four hexagon mats as shown in Figure 11.1f.

Figure 11.1f

Task 11.1.2 Isosceles Tetrahedron

Make a tetrahedron using four identical acute angle isosceles triangles.

What you will make is a, not very, irregular tetrahedron. Mark the mid points of all six edges. Label the vertices A, B, C, D.

Consider the three lines, through the inside of the tetrahedron, that join the mid points of AB to CD, AC to BD and AD to BC, that is, the mid-points of pairs of opposite edges.

What relationships can you find concerning these three lines? Make some conjectures.

Comment

Did you think to consider a regular tetrahedron as a special case (not just isosceles but equilateral faces)? You are likely to have found yourself focusing on one of these lines in your imagination, and then trying to introduce a second. If you used a regular tetrahedron as well, you have two instances in which to check any relationship to see if it is a potential property in general. You may even be able to imagine what happens if one vertex moves slightly: are the properties preserved?

There is a transformation proof that all three lines meet in a point, whatever the shape of the tetrahedron. It makes use of the stella octangular you made in Chapter 10.

Below is a word presentation of that reasoning. As you read, watch how you use your mental imagery, forming pictures or otherwise 'seeing' what is being described. (A word of caution; if you skipped making the stella octangular in Chapter 10, this will be a hard visualisation exercise. Perhaps you should stop and complete Task 10.3.4 before reading on.)

The vertices of a stella octangular are at the vertices of a cube. For any cube, the joins of the mid-points of the opposite faces pass through the centre of the cube. The centres of the six faces of the cube are the mid-points of the six sides of the tetrahedron. So these three lines also meet at the centre of the cube. Therefore, in a regular tetrahedron the joins of the mid-points of the opposite sides meet in a point.

Now, in your head, stretch the cube to form a cuboid (Figure 11.1g). This cuboid will contain four copies of a stretched tetrahedron around a deformed octahedron. More importantly you can see that your tetrahedron made from four isosceles triangles will fit inside a cuboid with the six tetrahedron edges lying along the diagonals of the six faces of the cuboid. The mid-points of each side of the tetrahedron therefore lie at the centres of the six faces of the cuboid. The joins of these opposite face mid-points can now be seen to meet at the centre of the cuboid.

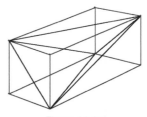

Figure 11.1g

Task 11.1.3 Irregular Tetrahedron

Extend the argument above by deforming the cuboid to become a parallelepiped around an irregular tetrahedron to convince yourself that it too has the property that the joins of the mid-points of the opposite sides meet in a point.

Comment

This result appeared as a problem in 2004 on a geometry discussion website. At one point the work was brought to a complete standstill by a contribution: 'This is a fact that every crystallographer knows'. It took some time for the mathematicians, who did not know it as a fact, to get back to working on the problem. People with different propensities (differently developed strengths) came up with reasoning based in different approaches: Euclidean, measurement, co-ordinate, transformation and others.

When you become aware that you are stuck on a geometrical problem, it is usual to try using familiar and confidence-inspiring tools (including dynamic geometry software, physical objects, co-ordinate calculations and so on), to specialise in some way (here, to move to a regular tetrahedron), to look for something familiar (the surrounding cube), and to seek some way to present the situation so that you can manipulate it confidently in order to get a sense of what is going on and what might be true in general.

All these strategies amount in the end to the same thing: to discerning and focusing attention on sub-figures, seeking relationships, perceiving properties, and reasoning on the basis of agreed properties. The teacher's task is to prompt learners to use and develop their powers (especially those powers which they are not yet inclined to use). This is done by directing learners' attention, and by using questions and prompts which support learners in shifting from discerning to seeking relationships, and from seeking relationships to seeing those relationships as potential properties.

Reflection 11.1

In what ways did the work in three dimensions alter or influence your perception of two dimensions? What do you do when you get stuck on a problem?

11.2 DEVELOPING CHAINS OF REASONING

This section considers the process of developing chains of reasoning, and comparing reasoning using transformations with other bases of reasoning. It begins by looking at centres of triangles, before settling on the centroid.

(Some) Centres of a Triangle

This section looks at just a few of the hundreds of points associated geometrically with a triangle, as a context in which to experience transformation-based reasoning and to compare it with Euclidean and measurement-based reasoning.

Task 11.2.1 Centres of a Triangle

For a scalene triangle, list as many different centres as you can. Mark each one on a different copy of the drawings in Figure 11.2a.

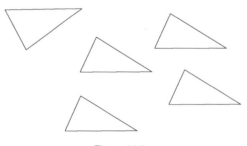

Figure 11.2a

Comment
This task, at a first level, might have caused you to question the use of the word centre. Surely, you could reason, there is only one centre of a triangle. What does centre mean? If there is more than one answer, then is centre an appropriate word?

'Centre' implies middle and perhaps symmetry, but scalene triangles have no symmetry! What all triangles do have is three sides and three vertices. *Tri-angle* stresses angles; an alternative would be *tri-gon* (*gon* for the Greek *gonos* for 'angled'), or even *tri-lateral* (which is used to refer to the figure in which the edges are lines rather than line segments, from the Latin *lateral* for 'side'). *Quadrilaterals* ought to have four (infinite) lines, while the more familiar figure ought (for consistency) to be called a *quad-gon* or *quadrigon*. *Quadrangles* ought to have four angles, but the term is now used to refer to the figure made up of six lines joining each pair from a total of four points. It is also used architecturally to refer to a space surrounded on four sides by buildings. With shapes that have more than four sides, English labels them by the number of vertices: pentagon, hexagon, and so on

In a triangle, things appear in threes. Centres arise by taking three attributes from a triangle and seeing whether they intersect in a common point.

Circumcentre: centre of the circumcircle, which is equidistant from all three vertices and is the intersection of the right bisectors of the sides.

In-centre: centre of the in-circle, which is interior to the triangle and equidistant from all three sides of the triangle, and which is the intersection of the angle bisectors

Ex-centres: centres of ex-circles, exterior to the triangle and equidistant from all three sides of the triangle, and which are the intersections of one interior angle bisector and two exterior angle bisectors.

Centroid: centre of gravity of the triangle thought of as made from a thin piece of uniform material; intersection of the medians (the lines joining each vertex to the mid-point of the opposite sides).

Orthocentre: intersection of the altitudes (lines through one vertex perpendicular to the opposite side; from *ortho* meaning right or normal).

For a taste of how the number of relationships explodes:

Nine-point centre: the centre of the circle which passes through nine special points: the feet of the altitudes, the mid-points of the sides, and the mid-points of the segments between the orthocentre and the three vertices. It was discovered by Jean-Victor Poncelet (1788–1867). The centre of this circle is the mid-point of the orthocentre and the circumcentre, (and the centroid also lies on this same line, known as the Euler line).

The fact that the three perpendicular bisectors of the sides of a triangle meet in a common point also appeared in Chapter 3. The fact that the three internal angle bisectors meet at a common point was developed from imaginings in Chapter 7. The fact that the altitudes intersect in a common point is proved later in this section.

Attention now focuses on one of the triangle centres, the centroid, and works on it in some detail. In the following tasks, as well as working at the mathematics, watch yourself working. What do you notice? How do you come to notice a particular property or relationship? How do you shift your attention from the first thing you notice to look for other possible attributes or relationships? The tasks are described in terms of using dynamic geometry software; an alternative would be mental imagery supplemented by a diagram when necessary, or lots of diagrams. Diagrams are omitted in the task in order to prompt you to draw your own.

Task 11.2.2a Centroids

In dynamic geometry software, mark 3 points A, B, C and use line segments to draw the triangle ABC. Mark in the medians, AD, BE, CF. Drag A (or B or C) around: what do you notice?

Comment

One striking feature is that the three medians always intersect in a common point. Some people are not impressed: they need to contrast this invariance with relationships which are not invariant, in order to appreciate the fact that the relationship is a property worthy of note. The interactive file '**11a Ratio-divided Sides**' shows what happens when the sides are each divided in the same ratio, for different ratios.

You may have found yourself wondering what dimensions of possible variation are available for the location of the intersection point: could it be at or at least near one of the vertices? Could it be outside the triangle?

Task 11.2.2b Centroids again

Denote the centroid (the centre of gravity) by G. Mark the mid-points of AG, BG and CG. What relationships strike you as you drag the vertices now? Join up those mid-points to form a second triangle. Does that reveal or reinforce any further relationships? Must the centroid of a figure always be interior to that figure?

Comment

It is much easier to notice that the medians meet in a point than it is to notice that they tri-sect each other. Marking the mid-points is an intervention which is likely to direct attention to the equality of the three segments. Equality is easier to notice than a ratio of even 1:2 between two moving lengths. Was the suggested construction a sufficient nudge? Were you able to shift your attention from the property of co-linearity to looking for relationships between lengths?

Did you notice any smaller triangles that were equal in area? If not, gaze again while thinking about area.

Drawing in the inner triangle was another way of prompting a shift of attention away from one property to seeking other relationships.

People often assume that the centroid lies inside the figure, but an annulus has its centre not inside the annulus, and this is the mechanism which enables tightrope walkers carrying a long pole to be relatively stable: the pole lowers their centre of gravity!

You can rarely predict where and how someone else's attention will be focused, which is why getting people to 'say what they see' is a useful pedagogic device to use with both small and large groups. What one person describes may require a significant shift in attention, and may reveal relationships not previously noticed. Saying to yourself what you can see is a useful internalisation of the same strategy, akin to writing down 'what I know' and 'what I want'.

You probably have two conjectures:

Conjecture 1: The medians AD, BE, CF intersect in a point G.
Conjecture 2: G lies at a point of trisection of the medians: AG = 2GD and so on.

These are still, for the moment, conjectures. There are three chains of reasoning given below, using first transformations, then a Euclidean approach and then one based on measurements and material world experience. They provide another opportunity to compare and contrast not only how you respond to different approaches, but to experience what it might be like for people whose preferences and predispositions lead them to prefer approaches with which you are less comfortable.

Chain of Transformation-Based Reasoning

The plan is to establish the truth of the two conjectures about a general triangle ABC.

Since properties are hard to deal with, start by working on conjecture 2 (a relationship).

It is required to show that G lies at the trisection of the medians: BG = 2GE and so on (restating the problem to yourself can often help concentrate the mind).

Since conjecture 1 is not yet established, delete the median from A. Gazing at your figure now it might occur to you to discern an absent feature: the line segment joining the midpoints of AB and AC, for example. Label them F and E respectively.

You may want to draw your own diagram.

Triangle AFE looks very much like triangle ABC: consider A as a centre of enlargement. Triangle AFE can be transformed (enlarged) onto triangle ABC using A as centre with a scale factor of 2. (Shift your attention momentarily to see why).

The effect of this enlargement is to show that BC = 2 × FE (all lengths are increased by the same scale factor). Now shift attention to BC and FE. There is another point which could act as a centre of enlargement sending FE to BC.

With centre G, triangle GFE will cover triangle GCB by enlargement, with a scale factor of ⁻2 since BC = 2 × FE

Hence GC = 2 × FG and GB = 2 × GE

Again a diagram will help if you are finding it hard to hold everything in your mind. You can use the previous diagram and superimpose your images on it, rather than drawing everything.

Put another way, the two particular medians CF and BE trisect each other at their point of intersection.

This provides a starting point to work on Conjecture 1: the medians AD, BE, CF intersect in a point G, by using the new information about trisections.

Task 11.2.3 Medians of a Triangle: Transformation Reasoning

In Figure 11.2b, use G as a centre of enlargement to map DEF onto ABC by enlargement.

Note the scale factor, and the image of GD under the enlargement.

What is wrong with this as a chain of reasoning? _____

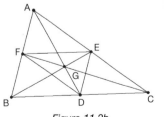

Figure 11.2b

Comment
The reasoning starts by assuming that G is the intersection of all three medians, and that there is an enlargement scale which takes triangle DEF to triangle ABC.

This demonstrates how you need to be careful to be aware of assumptions, to be self-sceptical in order to be able to convince someone else who is sceptical. Learning to question other people's reasoning is a good way to improve your own reasoning.

Chain of Euclidean Reasoning

Start with triangle ABC, as shown in the left diagram of Figure 11.2c. Construct the mid-points H of BG and I of CG, and join them with a segment.

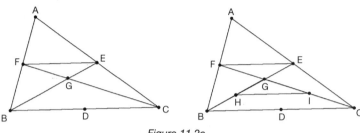

Figure 11.2c

Gazing at this diagram, and remembering that FE is parallel to BC (a Euclidean justification was given in Chapter 3), the two triangles EFG and HIG are strikingly similar, and potentially congruent.

Since HI is parallel to BC (again by the mid-point result), HI and FE are parallel. Some angle-chasing using parallel lines shows that the two triangles have the same angles and so are similar. Notice that you have to see one of them as 'flipped over' because H corresponds with E and I with F.

The triangles are actually congruent because both FE and HI are half of BC.

Alternatively, FEIH is a parallelogram (opposite sides being equal and parallel is a defining property for parallelograms). In a parallelogram the diagonals bisect each other (proved in Chapter 2 using congruent triangles).

Finally, since H is the mid-point of BG and HG = EG, G and H trisect the median. This reasoning applies to all the medians.

This reasoning demonstrates vividly the way in which a previous result is used as a property for reasoning: the segment joining the mid-points of two sides of a triangle is parallel to the base and half the length, and vice versa; diagonals of a parallelogram bisect each other.

Chain of Physics-Based Reasoning

In a science context the problem would be phrased as: consider a triangle of negligible mass and place an equal weight at each of the three corners. Find the centre of mass of the three weights.
   A science-based solution could proceed as follows:

The centre of mass of two weights of mass W placed at B and C will have the same effect as a single weight of mass 2W placed at D, the mid-point of BC (principle of moments).

The centre of mass of a weight of mass W at A and a weight of mass 2W at D will be a mass of 3W that divides AD in the ratio 1:2 (principle of moments).

Applying the same reasoning to other sides of the triangle, the three medians intersect at the centre of gravity which divides each of them in the ratio of one to two.

The physical meaning of centre of gravity can be used to justify the conjecture that the centre of gravity of any convex polygon must lie in the interior of the polygon. The reason is that, if not, placing it on its edge would make it tip over, and keep on tipping over, leading to perpetual motion!

In order to test your appreciation of the various forms of reasoning, you might wish to consider the analogous problem for a tetrahedron. Is it possible to modify the reasoning in each of the different approaches to cope with three dimensions (if, indeed, the result still holds!)?

To finish this section the next three tasks consider the orthocentre of a triangle.

Task 11.2.4a The Orthocentre

Imagine, or using dynamic geometry software, mark, three points A, B, C and join them up pair-wise using lines, not line segments. Draw lines from each vertex perpendicular to the opposite side. Drag the vertices around to get a sense of invariant relationships.

Label the intersection of two of the altitudes by H. What are the dimensions of possible variation in the position of H relative to the triangle? When is it at a vertex, or on a side of the triangle?

Comment
Again, surprise that the altitudes always (appear to) have a common intersection (an invariance) may depend on having had experiences where three lines do not always meet.

Task 11.2.4b The Orthocentre Again

Imagine a triangle A, B, C. Draw in the altitudes from A to BC and from B to AC. Let them meet at O. Then the line through OC is perpendicular to AB (in other words, the three altitudes of a triangle have a common intersection).

Draw and gaze at the diagram and see if you can justify the claim. Restate in your own words what you know and what you want to show.

Comment
The claim is deceptively simple, the possibilities are many, but finding a reasoning chain is actually surprisingly difficult. The next task suggests an approach using Euclidean circle theorems.

Task 11.2.4c The Orthocentre yet Again

Develop the diagram suggested by the following, and fill out the reasoning steps by inserting intermediate steps and justifying each one, in order to obtain a convincing proof:

Let the altitude from A meet BC at D, and the altitude from B meet AC at E, and let O be the point of intersection of the two altitudes.

The two right angles AEB and ADB mean that ABDE is a cyclic quadrilateral with diameter AB. The two right angles CE and CDO mean that CDOE is a cyclic quadrilateral with diameter CO.

> Extend CO to meet AB at F.
>
> Wanted: angle AFC to be a right angle. Gaze at triangle AFO. If its other two angles add up to 90° then AFC would indeed be a right angle.
>
> Use circle ABDE to find an angle at E equal to angle DAB.
>
> Use circle CDOE to find an angle at E equal to angle AOF (via 'vertically opposite angles').
>
> Conclude that since the sum of the two angles at E is 90°, so too is angle AFC.

Comment

Once seen, the reasoning is relatively simple, but there are no signposts to finding such a proof in the simple statement of the result.

Finally, here is an unexpected connection, which might or might not be the basis for a proof that the altitudes intersect in a point.

Task 11.2.4d The Orthocentre and the In-centre

Gaze at Figure 11.2d.

Triangle ABC was drawn first and then the in-centre and the three ex-centres were constructed. Consider the triangle $I_a I_b I_c$. Which of its 'centres' is the point I? Construct a chain of reasoning to show that the three altitudes of any triangle are concurrent.

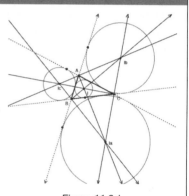

Figure 11.2d

Comment

The transformation you need to make here is of a different order. It is certainly not a reflection or an enlargement. You need to switch attention from triangle ABC as the core of the diagram to stressing triangle $I_a I_b I_c$ in relation to triangle ABC. For example, the sides of triangle $I_a I_b I_c$ are the bisectors of the external angles of triangle ABC; each vertex lies on an internal angle bisector of triangle ABC. But you know that the internal bisectors meet at a common point from the imaginings in Chapter 7. So the in-centre of ABC is the orthocentre of triangle $I_a I_b I_c$.

The important relationship is that for any triangle ABC there will always be a triangle $I_a I_b I_c$. A surprising relationship does not constitute a proof. What the comment says is that from triangle ABC you can produce another triangle $I_a I_b I_c$, but to prove that the altitudes of a triangle meet using this idea, you need to show that you can start with triangle IaIbIc and produce the triangle ABC, which is far from obvious. The next section looks at reversal constructions of this form.

Reflection 11.2

What is the same and what different about the reasoning chains to show that in each case (medians, angle bisectors, altitudes and right bisectors) the three lines have a common point of intersection?

What did you notice about what you had to do with your attention in each case? Recall now where your attention shifted between discerning features, seeking relationships, perceiving properties, and reasoning on the basis of properties?

Did you catch any moments when you were not sure whether a property was eligible for use in reasoning?

11.3 REVERSING CONSTRUCTIONS

Throughout mathematics the idea of reversing or undoing, that is, of interchanging the given and the deduced, the data and the answer, has proved to be remarkably fruitful. For example, the topic of group theory emerged as the study of this process. This section offers some examples of reversal or undoing tasks in geometry, associated with centres, starting with the reversal problem between orthocentres and in-centres.

Task 11.3.1 Orthocentre, In-centre and Ex-centres

Given a triangle ABC, construct a triangle PQR for which ABC is the triangle of ex-centres of PQR, thus showing that starting with triangle ABC, the altitudes must intersect (at the in-centre of PQR).

Comment
Usually the best thing to do in a construction task is to assume the construction completed, draw in the result, then look for relationships which might hold because of known properties. In this case, constructing the three altitudes without assuming a common intersection and constructing the triangle from the 'feet' of the altitudes on their opposite sides produces the expected triangle. The problem is to show that it is, indeed, the desired triangle.

To convince yourself that the construction works, it helps to make use of pairs of right angles which signal a circumscribing circle from which angle relations can be deduced.

Given a triangle, you can construct the three altitudes, angle bisectors and so on, and hence find the corresponding centres. Reversing the data and the given, suppose you are given three lines:

Task 11.3.2 Undoing Centres

Suppose someone presents you with three lines meeting at a common point, and claims that these are the altitudes, (or the right bisectors, or the medians, or the angle bisectors) of a triangle. Is there enough information to construct not just one triangle, but all possible triangles for which those three lines are the stated corresponding lines for the triangle?

For example, given three lines through a point, can they be the right bisectors for some triangle? What are all possible such triangles?

Comment
This question takes concentration!

It is typical of undoing questions that considerable creativity and insight is required. Where the 'doing' yields a single answer (for example, constructing the centroid of a triangle), the undoing questions often yield a family of solutions (for example, scaling a triangle from the centroid will not change the medians).

One approach is to act as if you have managed the construction: draw in the sought after triangle and then look for relationships. More detailed suggestions follow.

For each triple of lines, there has to be a common point of intersection. Using dynamic geometry software or pencil and paper, draw three lines through a common point. Gaze 'through the lens' of the properties those three lines are supposed to have (medians, altitudes and so on). Try a point as one vertex and see what this forces about the rest of the triangle. Then drag the point around until you get a triangle. Now find a property of the positions that seem to work so that you can write down instructions which construct them directly.

Construction Suggestions

Altitudes: pick a point on one altitude as the first vertex. This determines two edges of the triangle. The third edge is therefore forced. But is it necessarily perpendicular to the third altitude? Appeal to the proof that the altitudes must intersect in the orthocentre.

Perpendicular bisectors: starting from a random point P as a putative vertex, the other putative vertices can be found using the fact that the given lines are perpendicular bisectors. You might have become aware of a transformation, indeed a composition of three such transformations. Nominate the three given lines as mirrors M_1, M_2, M_3. Dragging the starting point P you eventually find a position where the last reflection point coincides with P. Notice that these special positions for P all lie on a straight line. Since the effect of reflecting in M_2 and then M_3 is a rotation through twice the angle between the mirrors (call it θ), you can work out a rotation of the first mirror through an angle of $180°-\theta$ to give one line of vertices. What is the effect of changing the order of the mirrors?

Angle bisectors: angle bisectors act as mirrors to send one side of a triangle onto another (even though the vertices do not coincide, the sides do). Label the angle bisectors M_1, M_2, M_3. Starting with a line L through a point P on M_1 (to serve as a putative edge of the triangle), reflect L in M_1 to give L' which is another putative side of the triangle. Reflecting L' in M_2 and then in M_3 rotates L' through twice the angle θ between M_2 and M_3. If L really is the side of the triangle, the result should be L again. So there is a relationship between the angle that L makes with M_1 and the angle θ, which enables you to construct L. An alternative approach is to use the relationship between in-centre and orthocentre, treating the given lines as altitudes first, then constructing the desired triangle from that.

Medians: picking a vertex on one of the medians, you know that the centroid C is $\frac{2}{3}$ of the way from the vertex to the opposite side, so the mid-point of the opposite side is known. Now the problem is to find a segment between two lines (the other two medians) which has a given point P as the mid-point. Gazing at this diagram for a while, it may occur to you that if you had such a segment, you would have a triangle CXY whose mid-points give lines parallel to the sides of the triangles (an appeal to Thales). This leads to the suggestion that lines through P parallel to the given lines will construct those mid points and hence, using Thales' theorem again, a construction of the segment with P as mid-point.

In each case some reasoning is required to justify the correctness of the construction.

Task 11.3.3 Variations on Undoing

What dimensions of possible variation occur to you in relation to the Undoing Centres task?

Comment
As well as working from centres, you could start with three lines as the lines joining the ex-centres and reconstruct the triangle. The variations start to multiply!

One response to the proliferation of exploration tasks in this section might be a sense of being overcome or overwhelmed by all the possibilities. Another response might be a welcoming of the opportunity to explore and to gain experience of gazing at a diagram and watching how your attention shifts as you work on the problems.

Reflection 11.3

What was your response to the proliferation of possibilities offered in these tasks?

11.4 MORE REASONING CHAINS

This section provides a final opportunity to work at locating and expressing chains of reasoning by flexibly moving between Euclidean, measurement and transformation-based perspectives.

More Examples of Transformation-Based Reasoning

The seven circle pattern in Chapter 1 is full of potential for developing chains of reasoning. For example, are the two rectangles shown in the diagram on the left of Figure 11.4a similar?

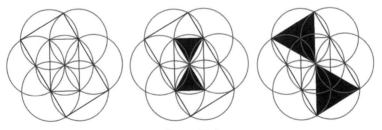

Figure 11.4a

At the time you may have solved the problem by drawing in the diagonals of each rectangle and noticing the head-to-head equilateral triangles sitting in each one as shown in the figures to the right of Figure 11.4a.

The aim of setting out reasoning chains, as initiated by Euclid, is to provide justification to convince any sceptic. Thinking in terms of transformations, the equilateral triangles support a sense that there is a rotation and a scaling involved. You could use a measurement perspective and compute, from the equilateral triangles, the ratio of the side lengths of both rectangles. Think transformations: gaze at the original diagram on the left of Figure 11.4a and imagine rotating the inner rectangle so as to align the sides with the larger one. In order to find out what the rotation angle should be, and what the following scale factor needs to be, you could set up the situation in dynamic geometry software, or start to calculate some lengths and angles. The prevalence of circles and equilateral triangles might suggest 30° or 60° and $\sqrt{3}$ as being involved. Some sort of manipulation in order to get a sense of the relationships is certainly called for. Once you have found a track, you can smooth out some of the dead ends and changes of direction to provide yourself and your audience with a clear chain of reasoning.

Task 11.4.1 Convincing?

Is this a convincing justification? Are you convinced? How would you articulate it so as to direct some-one else's attention? What objections could a sceptic raise?

Comment
Learning to be sceptical yourself is very useful in developing your powers to construct chains of reasoning which will convince others. By imagining objections and then address-ing them your reasoning will improve.

Extending Vecten's Diagram

The Vecten diagram (the second diagram in Figure 11.4b), which has appeared in sev-eral chapters, can be seen as part of a sequence.

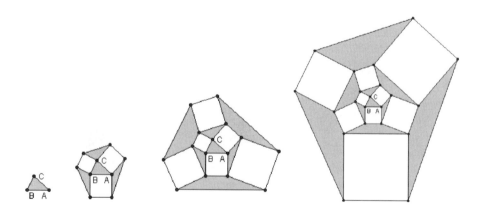

Figure 11.4b

The regions between the squares are suggestive in the figure: not only are they quadrilaterals but they look like trapeziums.

Task 11.4.2 Trapezia

Write down a rule for producing and extending the sequence of diagrams in Figure 11.4b. Prove that the quadrilaterals between the squares are trapeziums.

Comment
Recall or look up how the four triangles were shown to be equal in area in Chapter 7. Now gaze at one of the quadrilaterals and 'discern' an extra edge which would enable you to use the same idea as in the reasoning about the areas of the triangles. Still gazing at the same quadrilateral, 'discern' another extra edge.

  If you have two triangles on the same base with the same area, what can you say about the quadrilateral which contains them?

Drawing in the diagonals of the quadrilaterals reveals more 'triangles between squares' from which the same reasoning as before shows the triangles to be of equal area. But two triangles with the same area on the same base must have the same altitudes, so the base and the line joining the other vertices must be parallel, justifying the description of the quadrilaterals as trapeziums.

Having used the same idea several times it is often worth while isolating exactly what is being used so as to be able to use it as a property in future. In this case, at each stage only part of the diagram is being used at any one time: two squares with a common vertex and the triangles which lie between them, as shown in Figure 11.4c.

Figure 11.4c

The fact that two triangles have equal area can be seen in two different ways: using a measurement perspective as in Chapter 6 (area $= 1/2\ bc\sin A$ and $\sin A = \sin (180 - A)$) and using a transformation perspective (rotate one of the triangles through 90° about their common vertex which produces a single triangle and its median, so the areas of each half are the same).

This fact can now be treated as a property of hexagons made from two squares. In succeeding layers, discerning a diagonal of each trapezium, drawing it in and discerning two triangles reveals in turn, more such square-constructed hexagons. The same reasoning will apply to layer after layer, so at each stage the new quadrilaterals will be trapeziums.

Reflection 11.4

What triggers you to shift from thinking about and looking for transformations, incidence relations (congruence, circle theorems) and using trigonometry or at least ratio?

11.5 PEDAGOGIC PERSPECTIVES

Reflection on Reasoning

There is a problem when a proof depends on a neat preliminary construction, which is not always that obvious. Such constructions can appear to have been produced as if by magic. Pupils get no sense of ownership, no sense that they could initiate such an apparently disconnected starting point. Watching others articulate relationships and call upon properties does not guarantee that they too are discerning the same features, seeing the same relationships or perceiving the same properties. Somehow the initiative has to pass to the learner. Getting others to explain is really about having your attention directed appropriately, but you need then to take charge of directing your own attention if the experience is going to inform your actions in the future.

Reflection 11.2 invited you to review how you entered and worked on the preceding proofs. Perhaps you noticed that the presentations asked you to start by looking at a moving image, or a still image that, hopefully, invoked a moving image and start to consider it analytically. In each case the 'analytic' procedure is to work backwards to help you decide how to start. At some moment, with success in view, you disown all that preliminary stuff and announce your 'synthetic' proof as if it came pristine out of your head in the first place. In this way you strive to meet criteria set out by G.H. Hardy (1941, p. 29) when he remarked that in a good proof, 'there is a very high degree of *unexpectedness*, combined with *inevitability* and *economy*. The arguments take so odd and surprising a form; the weapons used seem so childishly simple when compared with the far-reaching results; but there is no escape from the conclusions.' He was talking about the finished product, the display of pyrotechnics that dazzle. He was not talking about the time taken to gaze at a figure, the false starts, the blind alleys, the time needed to take in other people's conjectures and synthesise those within one's own tenuously held set of connections.

Research into the development of reasoning (Hoyles and Healy, 1997; Hoyles and Jones, 1998; Hoyles et al., 2003a; 2003b) suggests that most learners make very little progress through secondary school in the forms of reasoning they can produce. If the stated aim is to produce chains of reasoning that get close to meeting G.H. Hardy's criteria it is not surprising that most learners 'make very little progress'.

If the aim is for learners to become confident and creative in their handling of visual images and reasoning with moving images then it is worth thinking about ways you worked on the median activities and how you, as a teacher, might work in this way with a group of pupils and other geometric content.

Why Is Proof Taught in Mathematics, and Particularly in Geometry?

Proof (chains of reasoning starting from accepted foundations) establishes mathematical facts as being unarguably true. The reasoning makes use of all explicitly stated assumptions, and in an ideal world, all assumptions would be evident. Thus mathematical reasoning explicitly specifies the range of permissible change of all possible dimensions of variation which make up the 'fact' (theorem). In order to appreciate any mathematical fact it is vital to bring to the surface the dimensions of possible variation and the associated ranges of permissible change of each dimension. This is what distinguishes mathematical reasoning from other forms of reasoning. Proof, the culmination of a successfully constructed chain of reasoning, is regarded as one of the crown jewels of mathematics. It enables learners to shift from being dependent on a teacher as authority for what is correct, to trusting mathematical reasoning as their authority. This is an important transition for adolescents who naturally struggle between wanting imposed authority and wanting to be independent of authority. Mathematics is one place where independence is possible. Of course, it helps to be inspired and guided by someone more expert, and to pick up ways of thinking and perceiving, and ways of reasoning, and clever insights that individuals might not come to. But ultimately the learner can be the authority. In this way, proof is as important in algebra and arithmetic as in geometry even though it is often seen, by outsiders, as sitting exclusively in geometry.

However, creating proof in mathematics is difficult. Many reasoning chains, such as encountered in this and previous chapters, seem to come from some great insight that is difficult to achieve. The challenge is to develop flexibility in approach, to shift from Euclidean incidence and congruence type arguments to ratios, and perhaps even trigonometry, to thinking in terms of transformations.

There is, of course, pleasure to be had from following a neat and concise argument. There is also pleasure to be had from gazing at a geometric figure, or an internally constructed image, and 'seeing' a relationship, whether a movement which could become a transformation, or a comparison which could become a ratio or trig calculation, or a potential congruence. The significant contribution to all learners' development available through geometrical thinking is to develop the power to imagine, to discern elements which are not shown, to 'see' a dynamic, as something is permitted to change, and to recognise that there are facts which must be true, relationships which may sometimes hold and relationships which can never hold. These facts and relationships are encountered and justified in the spartan world of geometrical diagrams, but apply to the material world. Architects, engineers, scientists and artists must have taken them into account in their professional activities. Working on an internal image as a way of seeing a result is a necessary precursor to being able to construct a logically connected chain of reasoning that could start to communicate that result to another.

Structure of Attention, Chains of Reasoning and Proof

Recall, and stay with a moment, when you gazed at a geometric figure, or an internally constructed image while working on a task in this chapter. Where was your attention? What sorts of things were you attending to, and when you made a step, how did what you attended to change? Thinking in terms of the structure of attention suggests that the shift from looking at the whole to looking at parts, from relationships to properties, and from properties to axioms and definitions, is so subtle and fleeting that it is hard to notice. Take this figure from Chapter 3. It is presented here (Figure 11.5) without the Chapter 3 starting question.

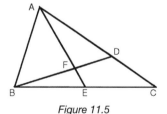

Figure 11.5

Task 11.5.1 Looking Again

Do not try to re-create what this was about in Chapter 3. Instead, gaze at Figure 11.5. What comes to mind?

What might you move (mentally), and what consequences would there be for other parts? Where did you first look?

Comment

Did you think of varying the shape of the triangle? Or the positions of D and E on their respective sides? You might have looked at the whole triangle, or at some part (discerning features). You might have wondered whether E is supposed to be the mid-point, or rather, taking control, you could decide to see what happens when E is the mid-point, or some other specified ratio along the side BC. Did you wonder if there is a relationship between the position of F along BD and the position of D along AC which might be independent of the specific triangle?

Did you think of areas? Did you think of extreme positions?

Each time you ask yourself one of these questions, you need to learn to pause and be explicit to yourself about the shifts you made, or pause and give learners time to articulate the relationships they recognise. Only after a respectable pause is it sensible to explore other objects to see if these have the same relationships. Make space, slow yourself down. Work on knowing a property or a relationship before moving to articulate the relationship as a property.

Train yourself to be explicit about the shifts possible in many different situations. Take time to expose relationships and to allow learners to express those relationships until their way of thinking is beginning to be in terms of properties. If you work at setting tasks which require a shift from testing whether a property holds to deducing facts from properties alone, learners' exposure to mathematical reasoning is likely to be more productive.

Your work on the above figure could have led you in all sorts of geometric directions. For instance: if instead of bisection of a side, you consider the trisection points on each side of a triangle, what might happen to the joins of the six lines (two from each vertex)? What relationships might there be if three lines each from a different vertex to the opposite side, all pass through the same point (this is *Ceva's theorem*). A great deal of work has been done on triangle centres, see, for instance; http://faculty.evansville.edu/ck6/tcenters and an associated, more algebraic site; http://faculty.evansville.edu/ck6/encyclopedia/part1.html. The first site is very accessible. The second one lists over 1,000 different triangle centres.

It does seem that lack of geometric content is not the problem. The problem is a pedagogic one; how can teachers stimulate learners to discover and gain access to this wonderful, varied and rich world of two-dimensional space? Since learners will pick up a great deal from the disposition and behaviour of their teachers, stimulating learners starts with being stimulated yourself. Noticing how you work, what happens to you as you tackle a challenge, sensitises you to what learners may be experiencing. Learning from your own experience can inform your decisions about how to guide and direct learner attention appropriately. There are no quick tricks or guaranteed strategies. The most you can do as a teacher is to be prepared to respond to your learners, informed by having experienced yourself what is possible. By focusing on the underlying purpose of geometrical tasks (not to memorise facts but to experience thinking and develop and extend learners' powers of mathematical sense-making), learners can get more from doing the tasks you offer them than simply filling in the blanks and copying out the facts.

Reflection 11.5

What have you learned in this chapter about how learners might be stimulated to discover and gain access to the wonderful, varied and rich world of two-dimensional space?

12 Visualising Transformations

This chapter starts by considering transformation as dynamic rather than static.

It goes on to look at composite rotations and reflections, and to show how in dynamic geometry software it is possible to construct the rotation of an object.

One of the powers of visualising with transformations is that attention is drawn to the intermediate positions between start and finish. This is the focus of the fourth section of the chapter. As before, the chapter ends with a pedagogic review.

12.1 MOVEMENT AND TRANSFORMATION IN ONE AND TWO DIMENSIONS

Transformations are often seen as static 'things' that take an input and spit out an output. Most geometrical transformations also have a dynamic aspect, for they can be seen as a movement in time and through space. In this section some of these movements will be used to generate phenomena in one and two dimensions. The next section will take up one of the ideas in three dimensions.

One Dimension

The section starts by asking you to think about transformations of a number line.

Task 12.1.1a Reflection on a Number Line

Imagine a number line, with 0, 1, 2, 3 and so on and ⁻1, ⁻2, ⁻3, ... marked on it. Think of this number line as fixed. Now imagine a second copy lying on top of the first, perhaps on an acetate sheet so it can be moved around. Imagine rotating the number line through 180° about the point 3.

Where has the point 5 on the acetate gone to on the fixed number line? The point ⁻5? Do enough examples until you can state where each point goes.

Comment
This rotation is also known as a *point reflection* because, just as with a mirror reflection, each point goes the same distance the other side of the mirror point. Did you think to generalise the role of the mirror point 3? Get a general formula for reflecting the point p in the mirror point m on a number line.

Did you think to consider special cases such as reflection in (rotating around) the point 0? There is a connection with multiplication by ⁻1. This in turn can be seen as a special case of scaling the number line by a scale factor of say s, where s takes on various values from 1 (no change) to other values including ⁻1. When $s = 0$ the number line collapses and disappears!

Task 12.1.1b Composite Reflections

Imagine a number line. Now imagine reflecting each point in the mirror point 3, followed by reflecting that in the mirror point 5 (the original 5 on the fixed number line). What is the overall effect?

Comment
Did you think to try some particular cases, or did you go immediately to the general, using your formula from the previous task? Is there a connection between reflection in a point on a number line and reflection in a mirror lines in the plane?

Reflection in a point can be thought of as reflection in a mirror line perpendicular to the number line and through the mirror point. But, more importantly, the composition of two reflections in parallel lines is a translation, and the composition of two reflections in mirror points is also a translation through a distance twice the separation between the points, and in the direction which moves from the first mirror point to the second. You can get a formula from the algebra which articulates this algebraically.

Before rushing on to the next task you might have paused to ask yourself whether the effect of a composition of three point reflections in three different points (say at m_1, m_2 and m_3) can be expressed more simply as a point reflection in some other mirror point related to m_1, m_2 and m_3.

The numbers on a number line are always considered to be uniformly spaced along the line. But by thinking about how the number line might be generated, other versions emerge.

Toppling Polygons

In this sub-section, you are invited to shift your attention from one dimension to two.

Task 12.1.2a Toppling Polygons

Imagine an equilateral triangle sitting with one vertex at 0 on a line, and another at 1. Imagine toppling the triangle over a vertex so that it lies back on the line. Keep toppling, backwards and forwards. The places where a vertex lands on the number line will be the other points marked with integer values.

Now imagine a scalene triangle (you could try it with a 3, 4, 5 triangle) sitting on the number line, one end at 0 and the other end at, well, in this case, 4. Now what happens as you topple the triangle over and mark the places where the vertices meet the number line?

Comment
There is a pattern but it is not entirely uniform. In the case of the 3, 4, 5, triangle which starts with the right angle at 0 and the other end of the 4-side at 4, you get a succession of segments of different lengths. The whole pattern repeats itself every 12 in either direction. Generalising to an arbitrary triangle is not difficult. Generalising to other polygons is also not difficult.

When generalising to other polygons did you think of a square or rectangle? What about a parallelogram, trapezium or kite, or even a pentagon (regular or not regular)?

What about the reverse or undoing problem?

Task 12.1.2b Toppling Polygons Reversed

If someone presents you with a pattern of segments tessellating the number line, what conditions must hold so that you can generate that pattern using a polygon?

Comment

A good first thought is that you 'just' make a polygon with successive sides having the required lengths until the pattern repeats. But maybe there is no such polygon. For example, a pattern of segments of lengths 1, 1, 4 repeating endlessly cannot be produced from a triangle.

The question actually comes down to what conditions must be imposed on a sequence of numbers so that these can be the lengths of the sides (in that order) for a polygon. The special case of a triangle suggests that the sum of the lengths of all but one side must exceed the length of the other side, otherwise the polygon will not close up. Thinking about this may have led you to wonder about non-convex polygons.

Very few people think about trying to topple a non-convex polygon. Indeed, it is not clear what it would actually mean. This is an excellent but simple example of how mathematicians often proceed. Since physically there might be a problem with toppling, you abstract the essential structure from the physical situation, ignoring physical constraints.

What is the essence of toppling? It could be seen as rotating the polygon around one vertex on the number line, through an angle which brings the next side of the polygon in contact with the number line, but always proceeding further in the same direction (that is, not back-tracking). Of course, you could use an alternative definition, which might be to rotate the polygon around the smallest angle (positive or negative) which will bring the next side of the polygon in contact with the number line. In this case, the points you mark on the number line might move backwards and forwards. Both 'definitions' are illustrated in Figure 12.1. The first three images show the quadrilateral toppling always to the right along the next available side, while the bottom row shows it toppling according to the smallest angle needed for the next side to coincide with the line.

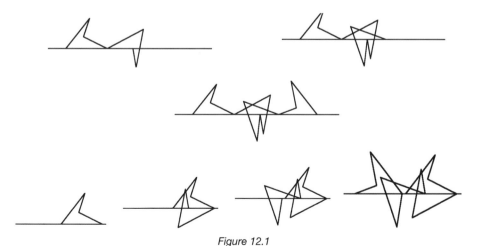

Figure 12.1

One possible direction for exploration would be to follow the track of one of the vertices as a polygon topples along a straight line. Its path would, of course, be made up of circular arcs. You could then ask what properties a path made of circular arcs would need to have in order to arise from the vertex of a toppling polygon. You could also track a point other than a vertex and study how these tracks change as the point being tracked varies inside the polygon.

Toppling is not an explicit mathematical topic on most school syllabuses, but it affords learners opportunities to make conjectures, to classify and characterise (what sequences of segments could come from a toppling polygon?), and to extend definitions to cover unexpected and unfamiliar variations (non-convex polygons). It also affords opportunity for learners to become aware of the strategy of using physical apparatus (to watch the toppling) and then to move to mental imagery and some other notation to keep track of what is happening.

Reflection 12.1

What did you notice about how you tackled the toppling tasks? Did you skim over them because they were unfamiliar? Did you use physical polygons or did you stick to mental imagery and the occasional diagram?

12.2 MOVEMENT AND TRANSFORMATION IN THREE DIMENSIONS

Before starting this section, it might be useful to construct a regular tetrahedron either by gluing together triangular mats or else folding a net. While you are at it, a cube would also be useful. An irregular tetrahedron made of four copies of the same scalene triangle would also be of assistance. There are two ways to do this, but one is a reflection in a plane of the other. A net can be made from parallelograms as shown in Figure 12.2.

Figure 12.2

Task 12.2.1a Toppling Tetrahedra

Imagine a regular tetrahedron. Try not to use a physical one at least until you have a sense of the full task! Now imagine the tetrahedron sitting on a large sheet of paper with its position marked as a triangle on the paper. Imagine toppling the tetrahedron over one edge, noting the outline of the face sitting on the base, and then toppling over other edges further and further. What sort of pattern will the footprints, the outlines of the faces, make where they sit on the paper?

Comment
It seems highly likely, if not pretty clear, that the tetrahedra will leave footprints of a tessellation of the plane by equilateral triangles.

In order to keep track of the positions of the tetrahedron as it topples, it is useful to label the vertices (1, 2, 3, 4 or A, B, C, D).

Task 12.2.1b Toppling Tetrahedra Again

Imagine (but you will probably need to do it physically) that you keep a record of which vertex was uppermost when the tetrahedron was sitting on any particular footprint, starting from some marked position. Since it makes a difference which way the tetrahedron is sitting on the starting footprint, you may want to record this somewhere as well. Since the tetrahedron can get to each footprint by a number of different routes, you may want to be systematic. From a starting position, which labels can be uppermost on which footprints?

Comment
Were you surprised? Did you think to label the vertices on the paper as well so that you could begin to do without the physical tetrahedron but begin to sense and then see a pattern in the labels of both vertices and footprints?

The task began with a regular tetrahedron, but you only partially understand something until you have considered dimensions of possible variation. What happens with an irregular tetrahedron? What could go wrong? Does it? There are opportunities for exploration here. You could also change the shape being toppled more radically.

Task 12.2.2a Toppling Cubes

Imagine a cube sitting on a large piece of paper. What is the pattern of footprints of the cube as it topples over its edges in all possible ways.

Comment
A square grid is a reasonable conjecture. Did you think to label the faces of the cube and record in each footprint the face which was uppermost?

You might have been surprised to find that more than one face can be uppermost on any particular footprint, yet there is a pattern to which faces can be uppermost in each footprint. There is a structural reason for what happens: as the cube topples round a vertex (so one vertex is invariant throughout), there are three faces which do not have that vertex. They appear successively in each of the four footprints round the vertex, and so eventually each face appears in each place. But now you can topple the cube in more complex ways to produce more labelling possibilities.

Once you see what is going on, you might like to pay attention to the orientation of each face when it is uppermost on any given footprint, perhaps by choosing labels which have no symmetries, so you can record their orientation on the footprint.

Task 12.2.2b Toppling Cuboids

If you place a cuboid on a piece of paper and sketch round the footprint, then topple it over on various edges by various routes, what sort of a pattern emerges?

Comment
Did you conjecture that you might get a grid of rectangles? Did the overlapping come as a surprise?

Experience with the labels on the cube might have alerted you to the possibility that the footprints would not form a nice grid because the three faces round a vertex all end up as footprints round that vertex. This raises the question of which three-dimensional shapes or solids will produce a simple grid when toppled over and over successively.

Reflection 12.2

How did your approach to the toppling tasks differ in three dimensions? Did you use physical objects this time or did you stick to mental imagery and the occasional diagram?

12.3 USING TRANSFORMATIONS IN DYNAMIC GEOMETRY SOFTWARE

In Chapter 9 it was pointed out that rotations and translations could be seen as the composite of two reflections. Translations are produced when the reflections are in parallel mirrors. The translation is through twice the distance between the mirrors and, in the case of rotations, through twice the angle between the mirrors. In both cases the direction is determined by the order in which the two mirrors are used. In this section this idea is used to analyse the effects of composite rotations and reflections, and to show how in dynamic geometry software it is possible to construct the rotation of an object (point, triangle and so on) about a specified point and through a specified angle.

Recognising Composites

When you encounter a complicated transformation that you want to classify (as a reflection, rotation, translation or glide reflection), there are two useful techniques. The first is to try particular cases using special points which simplify working out the effect of the transformation. The second is looking for the mid-point between a point and its image under the transformation. The reason for this is simple: for rotations, this point is invariant and there is only one; for reflections all these points lie on a straight line (the mirror) and for glide reflections they also lie on a straight line (the axis). Only for a translation is this mid-point unhelpful.

Task 12.3.1 Triangle-Based Rotations

Imagine a triangle ABC. Imagine an object (a point, a segment, a triangle, and so on) somewhere in the plane. Now imagine rotating the object first about A through the angle A (interior to the triangle in a clockwise direction), then about B through interior angle B, then about C through interior angle C. What is the effect on objects of this composite transformation?

Comment
There is an interactive file '**12a Tri-Rotations**' to support your work on this task. The object used in the file is a point P.

Did you try some particular cases (such as using an equilateral triangle)? Did you restrict attention to a single point, and relate its final position to its initial position? Did you sit back and think about, even imagine moving your body physically, to follow the rotations in order to experience the combined movement? Did you think to ask yourself what you know about the composition of rotations?

You might have noticed that the composition of three rotations must be a rotation about some point, and the angle of rotation must be the sum of the angles of rotation, which is a half a full revolution! So where is the centre of this combined rotation?

If you use the interactive file which performs the rotations for you, then you can find the centre by observing that it must be the mid-point between P and its image PABC under the composite rotations. As you adjust P you find that that mid-point remains invariant. Furthermore, it is reasonable to conjecture that the centre must be point on side AC of the triangle. Once you have a conjecture like that, it is often relatively straightforward to justify it. If you rotate the line through AC about A through angle A it ends up as the line AB, the next rotation takes it to the line BC and, finally, it comes back to AC except that it is in the opposite direction! In this case, think of performing the transformations on some point distance x from A, and see what happens.

> Let X be the point on AC distance x from A. Then rotating X about A through angle A anti-clockwise places it somewhere on side AB distance x from A and hence $c - x$ from B. Rotating this about B through angle B takes it to side BC a distance $c - x$ from B and hence $a - c + x$ from C. Finally, this goes to a point on AC a distance $b - a + c - x$ from A which should be the original point, at distance x from A. Consequently $x = (b - c + a)/2$.

Notice that repeating the whole composite twice results in a rotation through 360° and so takes every point back to its original position.

What happens if you alter the order of the rotations, or if you are not consistent in rotating in the same direction (clockwise) about each angle?

Rotating through 180° about a point is the same as reflecting in that point, which is known as a point reflection.

Task 12.3.2 Triangular Point Reflections

Imagine a triangle ABC and a point P. Reflect P in A (in other words, rotate through 180° about A), then reflect that in B then reflect that in C. What happens to P?

Comment
Thinking in terms of rotation in the plane may be easier. Three rotations through 180° means an overall rotation equivalent to a rotation through 180°. The question then is where the centre must be, and how it relates to triangle ABC. Often it helps to use special points such as the vertices of the triangle itself.

Completing the parallelogram ABC gives a fourth vertex D (CD is parallel to AB, DA is parallel to BC). You will need to convince yourself that this D point really is invariant under the composition of the three point reflections. Exploring what can be varied, you might wish to consider changes in the order of the point reflections (perhaps a different parallelogram is relevant). You could also consider point reflections in the vertices of a parallelogram, a square, or even a general quadrilateral.

Triangles have edges as well as vertices, which can be used to specify transformations.

Task 12.3.3 Tri-Reflections

Imagine a triangle ABC and a point P. Reflect P first in side AB, then reflect that in side BC and then that in side CA to give point Q. What is the overall effect on P? What is the overall effect if this whole procedure is applied again with Q as the starting point?

Comment

A dynamic geometry package can help, especially if you can put a trace on the mid-point between P and its final image under the composite transformation. See the interactive file '12b **Tri-Reflections**'. Casting about for special points such as when P is at B and when P is at a point whose first reflection is at C can be useful. Looking at the mid-point between P and its final image is also useful, especially in the special cases.

After some experimenting, the transformation begins to look like a glide reflection. Finding the axis requires considerable detective work. For example, taking P to be the point B as a special case helps because neither the first nor the second reflection affects this particular position for P. Thus the overall result for P at B is to send it to the reflection in AC. The mid-point between P and its image will be the foot of the altitude from B on AC. Trying P at point C is not as helpful, but choosing P to be the point which is sent to C under the first reflection is helpful, for the composite does not move it any more. This position for P is the reflection of C in AB, and the mid-point of P and its image is the foot of the altitude from C on AB.

The overall transformation is a glide reflection with the axis being the line through the feet of the altitudes from B and from C on the opposite sides. The translation is found by working out the image of P when it is, for example, the foot of the altitude from C. The image has to be on the axis just found and the same distance away from C (since for this special P the transformation is a rotation about C through twice the angle C). Recasting the transformation as this glide reflection reveals that when P wanders about, translating by the constructed vector and then reflecting in the axis achieves the same result as the three reflections.

Reflection 12.3

What did the tasks and the exposition offer you in this section?

12.4 TRACKING TRANSFORMATIONS

One of the powers of visualising with transformations is that attention is drawn to the intermediate positions between start and finish, whereas in a Euclidean approach attention is drawn only to two objects, and the relations between them (equality, perpendicularity, parallelism, congruence, similarity).

Tracking Translations

When you imagine a point it is not difficult to translate it in some direction through a specific distance and to keep track of all the positions it occupies on its journey. The result is a *line segment*, which represents one dimension. Instead of translating the point a finite distance, you could imagine translation infinitely far in one direction, resulting in a *ray*. Permitting translation both positive and negative, gives an infinite one-dimensional object, a *line*. There are difficult philosophical problems about what is meant by 'straight' or 'in one direction', but this is not the place to pursue these subtle problems. Making use of the same idea results in more dimensions.

Task 12.4.1a Moving Segment

Imagine your line segment. Translate it in some different direction (not along itself) through a specific distance. What do you get?

What special cases are available? What do you have to specify in order to get a rectangle? A square? A rhombus?

Comment

In general you get a parallelogram. The vertices are the start and finish positions of the ends of the segment, and the sides are the start and finish segments, and the track of movement of the vertices of the starting segment. If you used one- or two-way infinite translation you get infinite two-dimensional regions, including a whole plane if you permit full translations positive and negative.

To get a rectangle you have to specify that the second direction is perpendicular to the segment; to get a rhombus you have to specify that the distance travelled in the new direction is the same as the distance travelled by the point to create the segment.

Asking what special cases can arise and under what condition is an example both of specialising, and of exploring dimensions of possible variation in order to fully appreciate a concept or idea.

Task 12.4.1b Moving Parallelogram

Now imagine your parallelogram. Translate it through a specified distance in a direction which does not lie in the plane of the parallelogram. What do you get?

What special cases can you achieve and what do you have to specify in order to achieve them?

Comment

In general you get a *parallelepiped* (a squashed cuboid, all of whose faces are parallelograms). It is surprisingly difficult to draw a convincing parallelepiped because human perception tries hard to make sense in terms of a two-dimensional projection of a cuboid!

Why not continue the process? Imagine translating the parallelepiped in a direction not included already. The only difficulty is imagining 'where' that might be. But if you were a one-dimensional being you would already have had difficulty imagining 'where' a second dimension could be. If, instead, you let go of the need for a physical location and think purely in terms of 'a different direction not already available', then it is possible to convince yourself that you could carry on this process dimension after dimension. Of course it is difficult to draw, and so symbols become the easiest way to represent each step. Movement in one direction can be coded by a number: the distance travelled in the new direction. Each additional dimension requires an additional number to denote the distance in that new direction. For three dimensions you need three numbers, for four you need four, and so you find yourself thinking in terms of co-ordinates rather than spatially. Placing confidence in the representational system permits you to 'think' in as many dimensions as you wish!

Task 12.4.1c Moving Triangle

Start with a triangle and imagine translating it in a new direction (not in the plane of the triangle). What shape is produced? Replace the triangle with a square, or a circle, or some other shape.

Comment
All of the shapes share the property of being the translation of a bounded two dimensional shape, and so all are examples of prisms. Cylinders are then special cases of prisms.

In Chapter 9 it emerged that a translation could be thought of as a special case of a rotation, since a rotation is the composite of two reflections, and a translation is the composite of two reflections in parallel mirrors. This means that the preceding imaginings can be repeated using rotations instead of translations.

Task 12.4.2 Tracking Rotations

Imagine a point. Imagine rotating it about some different specified point through some given angle. Track that movement. Of course you get the arc of a circle, as a generalisation of a line segment.

Now imagine rotating that arc about some other point through some angle. If the new centre of rotation is in the same plane as the first arc, then you get a region in the plane. If the new centre is in some other 'direction', then you get a surface bounded by circular arcs, which you may be able to imagine by pulling the two-dimensional picture up into three dimensions (Figure 12.4). What properties does the region have analogous to the properties of a parallelogram?

Figure 12.4

Now imagine rotating the planar region through some angle about a new centre not in the plane of the arc. You get a solid region again bounded by circular arcs. If you rotate the surface through some angle about a new centre not in the space of the surface, you get a three-surface in four dimensional space! Hard to imagine perhaps, but progress can be made by careful sketching and calm imagining using those sketches. It is possible to make use of your physical 'sense' of rotation to get some sort of sense of the final object.

Comment
Compared with a parallelogram, instead of straight segments for edges you have circular arcs. Opposite arcs are equal in length and in radius (so they curve the same amount). However, most arcs lying on the surface formed, such as those acting as diagonals, will not be circular. Working out what 'shortest distance' means on surfaces even as apparently simple as these is quite difficult and a job best done by algebra.

Tracking Translations with Scaling

Once you think of translation as taking place in time, there is the matter of speed to consider. Here it is useful to assume that the translation is always performed at some constant speed. Suppose that, at the same time that something is translating; it is also changing size at a uniform rate, that is, scaling uniformly. What shapes will be swept out in this case?

Task 12.4.3a Tracking Scaling and Translating Together
Imagine a segment. Translate it in some new direction, but as it translates (at a constant speed!) imagine it being scaled at a fixed rate about its mid-point. What shaped regions can you fill out?
What conditions must be imposed to produce a trapezium? An isosceles trapezium (non-parallel edges are equal)? What happens if the scaling is performed about some different point (think Thales!)?
Comment At the start, the scale factor is 1. As the scaling increases at uniform rate, the length of the segment changes. Almost certainly the scaling you imagined was rather gentle, perhaps doubling or halving the length during the translation through a specified distance. You will always get a trapezium, because the final edge is parallel to the starting edge. If your scaling is more extreme, you find yourself discovering unfamiliar figures!

To understand an idea, such as the generation of shapes as tracking transformations, it is useful to consider dimensions of possible variation, and corresponding to each feature which can be varied, to ask what range of change is permissible. Here the change of the scale factor has been restricted to positive values. What happens if this restriction is removed?

Task 12.4.3b More Extreme Tracking Scaling and Translating Together
Now be a bit more adventurous. Allow the scaling to change from 1 to ⁻1 during the translation. What shape is produced? What conditions have to be imposed to see a triangle as the translation-scaling of a segment?
Comment The shape looks a bit like a bow tie. Its boundary is a quadrilateral, indeed a trapezium, but two of the edges cross. The point where the two edges cross has no special status and is certainly not a 'vertex', because it is not one of the ends of the segment at either the start or the finish position. To get a triangle, the scaling goes from 1 at the start, to 0 at the end.

This is typical mathematical behaviour: explore dimensions of possible variation and corresponding ranges of permissible change (in this case, allow the scaling to go through 0 and become negative), and include these under the definition. So more shapes count as trapeziums than you might initially have realised.

Task 12.4.4a Into the Next Dimension

Imagine a triangle. Translate it (at uniform speed) in a new direction (not in the plane of the triangle) but allow the scale to change uniformly at the same time. What shapes are produced?

Allow the scaling to go from 1 to 0. Allow it to go from 1 to ⁻1.

Comment

A possible name for this might be a bow-tet(rahedron) in the same way that a bowtie is a bow-tri(angle). In fact all of these shapes are *frustra*: portions of a cone, in this case, a triangular-based cone, which is a generalisation of a pyramid. The cone can be thought of as infinite if the translation and scaling are permitted to continue infinitely in both directions.

Task 12.4.4b Further into the Next Dimension

Imagine a square. Translate it in a new direction. The result is the frustrum of a (square-based) cone. When the scaling goes from 1 to 0 it is known as a (square-based) pyramid, which is half an octahedron. To see the octahedron, allow the scaling to go from 1 to 0 in one direction of translation, and from 1 to 0 in the opposite direction of translation. (Again, think Cavalieri: the solids produced all have square sections parallel to the base.)

Imagine a rectangle (not square). Translate it in a new direction but as it moves, allow the ratio of the sides to change uniformly from some value such as 2:1, through 1:1, to 1:2. Draw a net for the resulting shape. How might you describe this shape to someone?

Comment

The solid looks very much like a wedge. You might like to imagine it toppling over and over in different directions!

In the middle of the wedge there is a square cross section (when the ratio of the sides is 1:1), but this also happens in the case of a regular tetrahedron.

Task 12.4.5 Tetrahedron

Now imagine a regular tetrahedron. Find a cross-section which is square (mid-points of some edges is helpful). Now track what happens to the square as it is translated parallel to itself, but keeps its vertices on the edges of the tetrahedron. What shapes does it go through? How could these be described in terms of simultaneous transformations?

Comment

While the translation is taking place (at uniform speed), the ratios of the edges are changing at uniform speed, and the whole is being scaled at uniform speed, so that at the extreme positions you get the line segments forming two edges – those that were not swept out by the vertices as the shape was transformed.

There are, of course, many further questions to explore from looking for what can be varied, and how much it can vary. What happens if the tetrahedron you start with is not regular? What happens if you start with a cube, cuboid or parallelepiped: what sections can you achieve and what simultaneous transformations are taking place?

The kind of 'work' that you find yourself doing as you imagine, and push yourself beyond what is already familiar gives you some insight into what it might be like for learners when meeting ideas and ways of thinking which are familiar to you but not to them. Such experience acts as a reminder that teaching mathematics is essentially a process of prompting learners to make use of their own powers, and this you can best do by becoming ever more aware of how you use your own powers.

Reflection 12.4

What powers did you draw upon in order to tackle these tasks?

12.5 PEDAGOGIC PERSPECTIVES

Throughout this chapter, you will have been imagining, whether experienced as 'seeing pictures', as more of a 'developing verbal facility' or as more of a kinaesthetic experience. You were also invited to specialise (consider special cases) in order to appreciate the scope of the general. You may have found yourself conjecturing and trying to convince yourself, if not someone else.

There were opportunities to use your powers to specialise and generalise in trying to identify composite transformations. It might be the case that the extra experience of composing reflections to form a rotation has deepened your appreciation of the way in which reflections provide the building blocks for other isometries. It is most unreasonable to expect any learners to pick up an idea the first or even second time they encounter it. It often takes multiple exposures to the same idea, perhaps from different perspectives or in different contexts, for learners to begin to appreciate the richness and underlying structure of the idea.

Reflection

It is perhaps no coincidence that the transformations which generate all other isometries are reflections, and that the same word is used for thinking back over what has been done in order to learn from experience. Indeed, there is little evidence that people learn from experience alone. Rather, some sort of action upon previous actions is necessary in order for ideas and techniques to become robustly embedded in practice. Anne Watson (private communication) observed that one of the fundamental features of reflection is that it takes place in an extra dimension. To reflect one point in another on a number line you rotate the number line 180° about the reflection point; to reflect a point in a line you rotate the plane formed by the plane and the point through 180° in a new dimension; to reflect a point in a plane not containing that point, you rotate about the plane through 180° in a higher dimension. So, too, to reflect effectively on mathematical activity in order to learn, it is vital to step back and away, to move as it were into another dimension so that you can 'look down upon' the activity and see it as some sort of a whole. This is why a variety of types of task have been placed at the end of each section: the authors cannot predict what it is that you will need to reflect upon particularly, but it is possible to indicate the scope and range of reflection tasks which learners have found fruitful. So now is a good time to reflect upon reflection as a pedagogic or learning device.

How can learners be prompted to make effective use of reflection on their activity? Asking learners to construct a fresh example which illustrates what they have been doing has proved particularly effective (Watson and Mason, 2005). Inviting them to consider what powers they have used, what they had to do in order to make sense, can be alternated with asking them to express in words (or pictures or song or …) the essence of what a topic was about. They can also be asked to construct tasks which would be good for learners 'next year', or which would challenge another class or even one of the teachers. They could be invited to make posters which could be displayed at a parents' evening, through which parents could learn the mathematics for themselves. Attention could be drawn to aspects of the inner task which lie behind any task. Apart from the overt question or instruction as to what to do, there are mathematical themes which are likely to be encountered, mathematical powers which may come into play in some significant way, personal propensities which may emerge. Thus the learner may become aware of which of these are productive and which obstructive to their own efficient learning.

Perhaps the most valuable way to use reflection mathematically is to explore the dimensions of possible variation in tasks, because this is how rich appreciation of techniques and concepts develops. You only really appreciate the significance of a definition or concept if you are aware of what can change (and in what ways) and what cannot change in order to remain being an example. You only fully appreciate a technique if you are aware of what features must be present to make that technique useful. For example, when trying to recognise a complicated transformation, the technique of trying it out on particular points in special positions, and the technique of considering the mid-point between a point and its image are unlikely to come to you spontaneously unless you have used them successfully yourself in the past, and recognised them as effective techniques.

Notation

One of the affordances of the tasks in this chapter is the opportunity to experience how notation, even when it is just labelling, can be of tremendous help. A diagram can serve as the fixed element on the basis of which you can imagine and discern other elements or some elements changing, but it often helps to have some way to talk to yourself in order to refresh your image. Labelling vertices with letters (usually capitals) and using lower case letters for sides, with the side opposite a vertex using the vertex letter has become a convention because it facilitates thinking. In the toppling tasks, it really helps to note down the names of the vertices/edges where they meet the base recording sheet.

Tracking the locus of a point or segment as it moves is a form of notation which not only assists thinking about transformations, but leads to seeing how one dimension leads into two and then into three and then beyond, as well as raising a host of geometrical problems to think about and explore.

Reflection 12.5

What reflection tasks would you now set for each of the previous sections of this chapter?

Introduction to Block 4

This final block serves as summary, review, and reflection on the previous three blocks, with attention particularly on pedagogic implications. Here the mathematical powers, mathematical themes, pedagogic constructs and pedagogic strategies which have been used throughout the book are reviewed and summarised in one place.

Chapter 13 begins with some important themes which pervade mathematics and which serve to link and unify topics that otherwise might seem rather disparate. Even within geometry, the topics presented to learners often seem disconnected, largely because the encounter is based on mastering techniques without drawing on these themes or upon learners' powers. The remainder of the chapter considers the structure of mathematical topics, culminating in a framework which can be used to inform preparation for teaching any mathematical topic.

Chapter 14 begins with a description of some of the natural powers which all learners possess and which are vital for mathematical thinking generally, and particularly for geometric thinking. These are the powers which need to be activated and developed through work on specific topics, and you will have experienced them when working on tasks throughout the book. The chapter goes on to elaborate on the various pedagogic constructs and strategies made use of in and by tasks throughout the book.

13 Themes and Structure

Section 13.1 recapitulates some of the underlying mathematical themes which pervade and serve to unify mathematical topics. The next three sections each focus attention on an aspect of the human psyche – awareness, emotion and behaviour – as they impact upon learners' encounters with a mathematical topic. Section 13.2 draws attention to the sorts of things one would like to have come to learners' minds in the context of a topic. This is how learners educate their awareness. Section 13.3 considers motivational aspects of problem situations and applications associated with a topic which support learners in harnessing their emotions. Section 13.4 considers aspects of training behaviour in the context of a topic: gaining competence in procedures and methods. Section 13.5 then integrates these three aspects in a single framework which has proved useful for analysing and preparing to teach any mathematical topic.

13.1 MATHEMATICAL THEMES

There are certain themes which pervade mathematical topics, serving to reveal connections and links which might otherwise go unnoticed. The themes summarised here all have important manifestation in geometry, as will have emerged throughout the book.

Freedom and Constraint

A geometrical point has the freedom to be anywhere in space; two distinct points determine a line segment and its corresponding line, again free to be anywhere in space, with the segment of any length (other than zero); a third point at the mid-point of the first two is constrained by the choices made for the first two points, whereas a free choice of a fourth point somewhere on the line has some freedom, but it is constrained to be on the line. This trivial observation has far-reaching consequences. More complicated configurations of points, lines, triangles, circles and so on produce more sophisticated freedoms and constraints. For example, the Vecten diagram, versions of which appear in several chapters, starts with a triangle which has complete freedom (do you admit the three vertices to lie on the same line?), and then builds squares outwards on the sides of the triangle (Figure 13.1). These squares are constrained in size and position, unlike the triangle. The flanking triangles are then completely constrained: only by varying the original triangle can these

Figure 13.1

triangles change shape. Because of being completely constrained, all their features are determined by or depend upon the choices made in the original triangle. For example, the area of each of the flanking triangles has an invariant relation with the original triangle: their areas are all equal, and two altitudes of each flanking triangle are each equal to a corresponding altitude of the original triangle.

Dynamic geometry systems exploit this notion of freedom and constraint to offer an analogy between construction and axiomatic reasoning. Constructions depend on previous elements and so create dependencies or constraints. For example, starting with a triangle which is free, the medians are constrained, because they depend entirely on the choice of vertices of the triangle; alternatively you could construct a triangle by starting with three lines declared as the medians, and then see how much freedom there is in constructing a triangle. A more challenging task is to start with two intersecting line segments and construct a triangle for which these are two of the medians. Along the way you will find that the segments have to satisfy a constraint in order to be medians of the same triangle. Reasoning is based on agreed assumptions (often called axioms), analogous to the free choices in constructions. Subsequent deductions correspond to constraints. Thus, working with dynamic geometry software can afford learners opportunities to experience the structure of construction dependencies corresponding to the structure of chains of reasoning, where one result depends upon previously deduced results.

Most mathematical problems can be seen as starting from a very general, relatively free situation ('consider a shape in the plane or in three dimensions, ...'), and then imposing constraints upon that freedom. The problem is usually to construct one or more objects which satisfy all those constraints, or to prove that some further relationship must hold because of the imposed constraints (for example, that the medians all meet in a common point, or indeed the altitudes, the right bisectors, and so on). Sometimes it is necessary to demonstrate that the constraints are too demanding and that nothing can satisfy them all. Even the challenge of finding a proof (reasoning which justifies an assertion) can be seen as a constraint imposed on the freedom to deduce many different other results from the same starting points (there is a specified assertion to be justified and rules of logic to be followed, including giving reasons for every step). A useful way to get learners to become more aware of the choices and freedom available to them is to draw their attention to the freedom of choice when engaging them in some mental imagery (see Chapter 14).

Looking for freedom and constraint can be helpful in several ways. Stimulating learners to become aware of choices that they can make, and prompting learners to make choices, both contribute to a sense of involvement and participation which is the principal source of motivation. Seeing mathematics as a constructive and often creative activity makes a change from the more usual view of mathematics as a collection of techniques for solving predetermined problems.

Doing and Undoing

Constructing the medians, altitudes, angle bisectors, in-circle, ex-circles, circumcircle and so on from a given triangle can each be thought of as a 'doing': they are predetermined once the triangle is given. Each can be turned into an undoing: given three lines as the medians (or the altitudes or the right bisectors or the angle bisectors), construct all possible triangles; given a circle, construct all possible triangles which have it as the in-circle, the circumcircle and so on. Interchanging some of the 'givens' in a situation with some of the 'to finds' often produces challenging construction tasks which require changes in the way you think. While a 'doing' usually gives a single answer, corresponding 'undoing' often gives rise to a whole class of answers so there are dimensions of freedom to explore. Furthermore, very often there is some element of creativity, because more insight and ingenuity are required. Trying particular cases (specialising) in order to detect some underlying common structure (generalising) is often a good way to proceed.

Extending and Restricting

It is common for learners to see rectangles and squares as different, rather than to see squares as a special case of rectangles. A rhombus is a special case of a parallelogram, as is a rectangle and hence a square. Sometimes special cases are so special they challenge intuition: for example, if you start with a triangle and slowly move one vertex towards the opposite side, there is a point at which the three vertices all lie on the same straight line. This is known as a degenerate triangle, but a triangle nevertheless: the sum of its angles is 180°, the area formula correctly gives 0, and it satisfies all other theorems about triangles. It is usually useful to include all possible special cases rather than to rule them out. Thus a triangle is also a special case of a trapezium (one of the 'parallel sides' is of zero length), which in turn is a special case of a quadrilateral. The area formula for the trapezium is correct in the special case of the triangle, as is the area formula for the quadrilateral. Something is a special case of something else when all theorems for the more general remain true for the particular.

Seeing one named property as a special case of, or included in, some other property is a form of extending meaning from the intuitive and familiar to the general. Another direction for extending meaning in geometry is to take concepts on the plane and to reinterpret them in other spaces such as on the surface of a sphere or other surface, or to interpret different objects as points and lines (for example, taking as points the points inside a specified circle, and as lines, the circular arcs inside the specified circle which meet it at right angles; taking circles in the plane as 'points', and taking certain families of circles as 'lines').

Restricting meaning is required to focus attention only on certain objects such as all the points inside a specified circle and treating all others as simply not available or even not existing. A particular case where extending and restricting appear is in the theorem that the angles subtended by a segment in a circle are equal. There is a problem in interpretation when the angles are on opposite sides of the segment, but if you are very careful you can define what you mean by 'the angle subtended by a chord' so that the theorem remains true no matter which side of the segment the angle is subtended. This illustrates how mathematically it is useful to adjust definitions until they contribute to the succinct statement of theorems which apply to a general class of situations.

Extending meaning is closely related to considering the range of permissible change of a parameter or feature of a concept, task, shape, and so on. By inviting learners to consider explicitly what the range of permissable change might be, the teacher can help broaden considerably their appreciation of the scope and range of a technique or concept. Conversely, if learners have a narrow sense of the range of permissable change in a situation, they may not appreciate what the 'fuss' is about.

Invariance and Change

Mathematics is only peripherally concerned with very specific facts, such as that $\sin 30° = 0.5$. Rather, it is concerned with generalities, and these can often be approached by thinking about what is allowed to change, and what remains invariant in the midst of that change. For example, in $\sin^2 A + \cos^2 A = 1$, the presence of the letter A suggests that its value can change, but that the 1 is taken to be invariant. So, too, the fact that the rectangles formed within the squares in the Vecten diagram are equal in area is a theorem which applies to any triangle whatsoever. Indeed, it applies even when the square is changed to 'similar rectangles', and it is possible to reformulate it so that similar rectangles can be changed to 'similar parallelograms', and still the equality of areas remains invariant, as does the equality of the areas of the flanking triangles. With any geometrical result or theorem, it is useful to ask yourself, what is it that is permitted to change (dimensions of possible variation) and in what ways (associated range of permissable change), and what remains invariant. What remains invariant is usually a relationship. That relationship can then be treated as a property to be checked in other situations (see 'Structure of Attention' in Chapter 14).

Human senses (seeing, hearing, touching, smelling, tasting and perhaps a less choate 'sense of structure') all work by detecting change. But change only makes sense if there is something that is not changing (or not changing so much) as a background against which to detect the change. It is often useful to switch from stressing what is changing, to stressing what is staying the same and back again in order to appreciate the significance of a collection of objects or 'examples'.

In several tasks you have been asked to consider what was the same and what different about two or more 'objects'. The idea is to make use of your power to select attributes and to detect pattern or similarity. This leads to seeing particular objects as particular cases of a general class of such objects. Once the objects are seen as representative of a whole class of 'similar' objects, a phenomenon has been identified, based around that sameness. This is the essence of generalisation, seeing the general through the particular.

To appreciate sameness and difference requires that you discriminate features, some of which are shared by all objects being considered but which might not be shared by all possible objects (sameness), and some which can be used to distinguish between the objects (difference). In trying to transform a task they have been set into a task they can do, learners often stress features that are not what the teacher is stressing. The result is that they may find generalities being asserted by the teacher rather opaque!

When you stress certain features apparently shared by all the objects, and ignore differences, you are engaged in abstraction or generalisation. You are classifying according to those selected features, opening up the possibility of considering other objects with those same features, but perhaps differing in some other respects which are currently being ignored. As Caleb Gattegno (1987; see also Brown et al., undated) said, stressing and ignoring is the basis for generalisation and hence for classification.

Geometers are interested in finding relationships that hold across all possible examples, for all possible circles, then treating those relationships as properties to prove invariant. For example, many different tangential quadrilaterals can be constructed about a circle; but the sums of the lengths of opposite sides are equal.

Learning from and through Experience

Four mathematical themes have been put forward: Freedom and Constraint, Doing and Undoing, Extending and Restricting, and Invariance and Change. The claim is that these not only pervade mathematics, but often serve as links and connections between otherwise apparently disparate topics. Put another way, learning mathematics can be perceived as exploring the variety of ways in which these themes are developed and instantiated or exemplified in different mathematical concepts and topics. By being aware of these as themes you can draw upon them when structuring tasks, so that learners encounter the themes, and you can use the themes explicitly with learners as prompts to get them to make choices and to develop and extend their appreciation of concepts. For example, when stuck on a problem, learners can be invited to ask themselves what freedoms and constraints they are assuming, and whether they are unintentionally restricting or extending meaning; when they think they have finished a task they can be invited to ask themselves whether a 'doing' could be converted into an 'undoing'; when learners are being introduced to an extended meaning of a term with which they thought they were familiar, you can be sensitive to their surprise and confusion at finding old ideas being extended or varied.

13.2 AWARENESSES AND ABSENCES

Mathematical concepts and topics are not simply 'ideas', nor are they simply 'definitions' or techniques. Rather, the words are labels for a complex tapestry of interwoven thoughts and images, connections and links, behavioural practices and habits, emotions and excitements. When something becomes a topic to be taught in school, it is because someone recognised that a class of problems could be solved by using a technique, which depends on being aware of certain concepts, which are themselves ways of perceiving, ways of stressing some features and ignoring others. This and the next two sections introduce the notions of awareness, emotions and behaviour as components to the structure of a topic, which is then summarised in Section 13.5.

Awareness

The totality of thoughts, images, ideas, associations, related topics and concepts which come to mind when you encounter a word or an idea constitute your awareness in the moment.

Task 13.2.1 Coming to Mind

What comes to mind when you encounter the terms *Pythagoras' theorem*, or *sin A*?

Comment
Right-angled triangles are likely to feature, but are you aware of diagrams, of situations in which they have been used? Did you think about Pythagoras as a historical character, and were you triggered to think about other aspects of trigonometry? Were you aware of a connection between Pythagoras' theorem and the trigonometric identity that $\sin^2 A + \cos^2 A = 1$?

As thoughts come, other awarenesses may also come to the surface. These are the awarenesses which are dominant for you in association with the term or topic as triggered by a particular situation.

The important question to ask yourself is which of these you want to come to learners' minds as a result of work on the topic. Similarly, you can ask of any set of exercises what awarenesses are or could be coming to learners' minds as a result of working on the tasks. Thinking in this way orientates you towards choosing how to introduce, develop and complete work on each and every mathematical topic, and how to frame tasks so that learners get the most out of exercises.

A powerful way to think of awarenesses is tied up with dimensions of possible variation and ranges of permissible change: what are the dimensions of possible variation of which you are immediately aware, and which come to mind upon further thought? Do learners appreciate the range of permissible change that you are aware of in each dimension? If not, how might you stimulate them to extend those ranges, to appreciate the full generality necessary to appreciate the concept or topic?

Absence

Learners for whom some aspects of a concept or topic do not come to mind may be disempowered. Furthermore, some ideas that do come to mind are not always appropriate or even correctly formulated, so associated with *awarenesses* are *absences*: things that learners often forget about, misconceptions and misconstruals that you notice learners having to work their way through at various times, and classic errors which learners seem prone to making.

There are a number of classic confusions and absences that teachers and researchers have observed, such as:

- thinking that a shape is only a triangle when the base is horizontal (aligned with the bottom of the page);
- not seeing squares as kinds of rectangles, triangles as kinds of trapezia or quadrilaterals, and so on;
- associating size of angle with size of the arms rather than the rotation between the arms;
- mixing up conditions for congruence between triangles;
- 'knowing' that vertically opposite angles are equal, but not recognising angles as being 'vertically opposite', and even if they recognise them, not thinking of their equality as a property that all examples must necessarily possess;
- learners identifying shapes by global characteristics (circles are round) rather than by more technical properties (set of points a fixed distance from a fixed point);
- learners relying on visually apparent equality or other relationships when justifying conjectures about a figure;
- learners perceiving properties but not reasoning solely on the basis of agreed properties, but rather invoking everything they think they know about a shape or configuration;
- learners unaware of the scope and range of generality implicit in the statement of a result;
- learners under the impression that measurement (whether of objects or of diagrams) justifies assertion of equality or other relationships.

One of the issues concerning absences is whether these sorts of 'mistakes' are slips because attention is fully involved in some other aspect, or whether they represent incomplete or flawed sense making. Which they are will make a difference in how they are dealt with. In Section 13.4, ways of exploiting classic learner errors are proposed.

Most of the absences mentioned here can be accounted for by thinking in terms of the structure of attention (see Chapter 14), for they arise when learners are attending in one way but expected to attend in a different way.

Anthropologists have coined the expression 'absence of evidence is not evidence of absence'. In other words, just because a learner gives no evidence of being aware of or knowing how to do something, it does not follow that they are not aware or do not know how to do it. There may be other reasons why it did not come to mind, or was ruled out of consideration. Consequently, care must be taken not to ascribe absences to learners, but rather to their behaviour at a particular time. Asking learners to construct objects is the most effective way to get them to reveal what they are aware of (see section 14.3 for specific pedagogic strategies).

It can be very helpful, especially when preparing to teach a topic or when refreshing your sense of a topic before teaching it again, to have kept a notebook in which you recorded awarenesses (images, ways of thinking, links, sense of 'essence' of a topic and so on) that you noticed at different times, and errors, absences and confusions which you detected learners experiencing. The next two sections will suggest further pages for the notebook, so that in total you might want to have either a separate notebook for each topic, or several pages devoted to each topic. One pair of pages would be *awarenesses* and *absences*.

13.3 HARNESSING EMOTION

Every mathematical topic arises because someone works out how to solve a class of problems. The solution may involve new concepts, a new technique or a new way of thinking. A curriculum designer then decides that the concept and the technique are important and within the reach of learners, and so the topic is introduced. Unfortunately, what often happens is that the concept and the technique are all that remain when the topic is described in the curriculum. Yves Chevellard (1985, see Mason and Johnston–Wilder, 2004) introduced the notion of a *didactic transposition* to capture the change which takes place when an expert converts their awareness into instructional materials: what emerges is instruction in behaviour for learners. What is lost is not only the awareness, but the emotional (motivational) aspects which accompanied the expert's crystallising awareness of underlying connections and mathematical structure. All that remains is techniques and concepts. Access to the original situation which puzzled someone and led to the solution may be missing, yet this is where the potential motivation lies.

Encountering Surprise

Being surprised by something ignites interest. If that interest can be sustained so that it turns into a flame, then learners 'have been motivated'. Contrary to most learners' experience, every mathematical topic has an element of surprise in it, otherwise it would not be a topic. For example, the fact that the angle between the two lines joining a point to a chord on a circle stays the same when you move the point round the circle can be experienced as a (mild) surprise, leading to wanting to sort out why this happens. The fact that this angle is always half of the angle subtended by the same chord at the centre can also evince some surprise: why is it constant? Why is it *that* constant? What is the focus of points which subtend an angle one third the size of the angle at the centre? Are circles the only curves that have this property? This last question is an example of reversal, turning a doing (angles subtended at the centre are twice the angle subtended at the circumference) into an undoing (the locus of points which subtend half the angle subtended at a point is a circle with that point as its centre).

You were probably familiar with Pythagoras' theorem before you started this book, but you may have been surprised at the more general case arising from the Vecten diagram (Figure 13.3), including the equality of the areas of the flanking triangles. There are many surprises to be found within this diagram. For example, each flanking triangle can be constructed from the central triangle by cutting the central triangle in two along the median from the associated vertex, then rotating each part about the centre of the associated square. (See Figure 13.3.) Notice also that the median of the flanking triangle in the figure is perpendicular to the base of the original triangle, so it is an extension of the altitude.

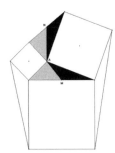

Figure 13.3

By experiencing surprise, you are in a better position to stimulate and motivate students to experience that surprise themselves.

The Role of Contexts

The emotions are the source of energy, of drive and initiative. Emotions need to be harnessed, that is, directed into focusing attention appropriately rather than dissipating it. The best way to do this is to invoke surprise and to expect learners to exercise their powers. It can help learners if they appreciate the sorts of problems that people encountered which are resolved by the topic, and the contexts in which that topic and its techniques are likely to arise. This is no mere window–dressing, with a few applications stuck on at the end which only the quick-working learners ever get to. It means assisting learners to relate the contexts in which typical problems arise to their own experience. It does not mean basing everything only on authentic problems which people outside school encounter (so called *authemtic mathematics*). It does not mean always starting from some practical situation (as advocated by *real problem-solving*). It does mean making contexts realistic (capable of being realised and appreciated) by learners (as advocated by the *Realistic Mathematics* perspective developed by Hans Freudenthal and colleagues at the Freudenthal Institute in the Netherlands). This may take the form of imagining some situation, making use of some authentic problematic situation, drawing on historical information or putting learners in some mathematical context in which a problem arises.

Some people like to have an overview before committing themselves to specific tasks; others are happy to engage immediately. It is usually best to cater to both dispositions by varying your approach, sometimes beginning with potentially realistic settings, and sometimes going straight for structure. Mathematics is well known for the use of 'motivating examples', but often these involve contexts which are anything but familiar to learners. At the other extreme, examples can be drawn as often as possible from everyday situations which learners recognise they are likely to meet. As long as the work on the topic is not confined to such contexts, it can help learners see why they are being asked to engage with the topic. Some teachers find that a strong motivating force is a sense of meeting and overcoming a challenge, through the use of learners' own powers. Others find that presenting learners with a phenomenon: something happening, perhaps on a screen, stimulates curiosity through generating surprise.

Using Powers

Human beings get pleasure from using their powers (summarised in Chapter 14). They often become frustrated when other people do things for them which they are on the edge of being able to do for themselves. Indeed, learners sometimes decide that their powers are not wanted in the mathematics classroom, and so they stop using them even where there is an opportunity.

Developing geometric thinking both exercises and develops those powers in everyone, particularly the pleasure of encompassing a wide range (even better an infinity) of possibilities under one expression of generality. Using their own powers in a supportive atmosphere also reveals to individuals the fact of those powers, something learners may not be aware of until it is drawn to their attention. There is nothing so strongly motivating as realising you can do something that is valued and valuable. The exercise of your own powers, independently, is a major source of pleasure for human beings, whereas dependency on others breeds discontent.

Learner Constructions

A rich domain for making choices is to get learners to construct mathematical objects for themselves. For example:

Construct a quadrilateral in which two sides are equal;
… in which two sides are equal, and two sides are parallel;
… in which two sides are equal, two sides are parallel and two opposite angles are equal;
… in which two sides are equal, two sides are parallel, two opposite angles are equal and two adjacent sides are perpendicular.

At each stage, as the constraints pile on, the aim is to find the most general class of objects rather than just a single example. In other words, do not be satisfied with a special case like a square or a rectangle for the first three, at least. It may turn out that all the constraints together are impossible, but this conjecture will require reasoning to justify it. It may be necessary to extend what you mean by the terms used (in this case *quadrilateral*) in order to permit a range of change that you had not previously considered (in this case, that the sides can cross).

Any problem which asks for a solution to some constraints can be seen as a construction task, where you start from no constraints and then gradually add the constraints on one at a time. In each case, you express the most general objects possible which satisfy the current constraints. By breaking down a complicated task you can make progress and exercise learners' powers to make mathematical sense. These ideas are developed in Watson and Mason (2005).

As well as asking learners to construct objects, you can ask them to 'say what they see' in geometrical diagrams and animations, for example, while dynamic geometry software is being used (see Chapter 14 for elaboration of this and related pedagogic strategies). Both kinds of tasks are useful for getting learners to reveal the scope and range of their awareness of generality in relation to a specified concept or property

Making Sense Mathematically

The world of the twenty-first century is full of situations which require some form of mathematical thinking in order to make sense of them. Chapters 4 and 8 offered some examples of these where geometric thinking is useful. Geometrical thinking applies to situations in the material world where some facts must always be true, others may or may not be true according to extra conditions and other things simply cannot ever happen. Emotions are evident when learners realise that a situation is a special case of something more general, or when several situations are revealed to be instances of some general unifying idea.

13.4 TRAINING BEHAVIOUR

To perfect a technique requires much more than simple repetition of behaviours until it is automated. Every topic involves technical terms and phrases or sentences which express relationships and properties. Becoming familiar with those technical terms means not only having a sense of what they mean, but actually making use of them in order to express your own thinking. Associated with every topic are collections of words, phrases and sentences which it is worth recording in a notebook as a reminder for the next time the topic is to be taught. Sometimes learners come up with humorous or interesting variations, which are also worth recording so that they can be used or circumvented in the future, as seems relevant. Often things that learners say are based on things they think they have heard the teacher say, so by reminding yourself, before starting a new topic, of some of the things that might derail or confuse learners, you can avoid some of the confusions that arose in the past.

Genuine discussion involves the negotiation of meaning between and among participants. The act of discussing, involving the sometimes tentative use of new words, is part of the process of arriving at meaning and understanding. This is part of maintaining a conjecturing atmosphere in which what is said, no matter by whom, is treated by everyone as a conjecture to be tested in mathematical reasoning.

Practice Makes ... ?

What is being practised when learners are set collections of exercises to complete? Are they actually developing facility in the use of particular concepts and techniques (that is, are they learning the ins and outs of particular concepts and the use of particular techniques)? Are they using and developing their powers? Encouraging learners to articulate how you do 'one like these' and to describe what makes something 'like these' contributes to their awareness of the topic rather than just working at training their use of techniques.

Geometric reasoning is a particularly difficult practice to develop in learners. They have to be in the presence of reasoning, when it matters to them, and they have to find themselves trying to justify their conjectures to others by constructing chains of reasoning, all over a period of months not just days.

Practising something overtly can be useful when you are trying to gain facility and fluency, that is, to automate procedures. But the most effective practice occurs when attention is directed away from the performance, so that the behaviour is subordinated, so that it no longer requires much explicit attention. As long as attention is directed towards the technique or behaviour you want to automate, the process of becoming automatic is delayed.

Learners' facility and fluency can be tested by giving them sample tasks to do in a limited time. Learners' awareness is best revealed by asking them to construct relevant objects meeting suitable constraints, and inviting them to push the boundaries of the dimensions of possible variation and associated ranges of permissible change of which they are aware.

Using Classic Errors

A useful source of explorations involves classic errors (absences) that learners have made in the past. By collecting these for each topic as suggested in section 13.2, you equip yourself to confront similar errors in the future. For example, scaling a shape by adding rather than by multiplying the side lengths comes up in all sorts of different contexts: adding is more natural than multiplying, due to the amount of time devoted to it in the early years.

Offering a counter-example may convince learners, but it may only be seen as a passing incident which does not actually challenge their conjecture, which itself may not actually be explicitly formulated. Getting the learners to articulate the conjecture and then to test it for themselves turns the initiative over to them, and reinforces the notion that in mathematics the authority lies within mathematics, not with individual people.

13.5 STRUCTURE OF A TOPIC

Every mathematical topic consists of three strands, corresponding to the previous three sections, because these reflect the structure of the human psyche: awareness, emotion and behaviour, or in more scholarly terms, cognitive, affective and enactive aspects of human psyche. When preparing to teach a topic, whether for the first time or when refreshing before teaching it 'yet again', it can help to review the three strands in order to be reminded of nuances and observations you have made in the past. Hence it helps to have a notebook with three pages for each topic: one page for each strand.

The awareness strand encompasses the ideas and thoughts that you find come to mind as you think about the topic, and hence that you would like to come to learners' minds as well. This includes mathematical themes, images, diagrams, connections with other topics, and the like. This strand also includes the classic errors that learners make, whether as slips when not paying sufficient attention, or more deeply based absences or misconstruals of what they have encountered. These are worth collecting over the lifetime of your teaching, as they indicate ways in which learners attend to what they are presented with, which may not be what was intended, and so indicate places where further care might be valuable when teaching the topic next time.

The emotional or affective strand includes both the problematic situations which originally gave rise to the topic, together with other situations which learners might be able to appreciate, and also the range of other situations in which the topic or some aspect of it is likely to arise. This goes beyond 'applications', to include situations in which the ways of thinking have proved relevant. For example, in the context of Pythagoras, ancient Chinese manuscripts are full of problems about a reed in a lake which is broken, so that the tip just touches the water; lengths are given and the task is to find the distance of the tip from the stem, or to find the original length of the reed above water. This is of little practical use, yet nevertheless of historical importance and hence possible interest. An observation about adolescents is that claims to be 'bored' are often statements about lack of confidence rather than lack of motivation.

The most familiar aspect of a mathematical topic is the techniques or methods which learners are expected to master in order to solve test questions. But there is more to behaviour than methods. For example, every topic involves the use of special terms, phrases and sentences which both orientate and demonstrate relevant ways of thinking. Every technique or method is accompanied by 'things you say to yourself as you are doing them', perhaps reminding yourself about choices, or about tricky bits that you sometimes get wrong. These 'inner incantations' are just as much part of the behaviour associated with a topic as the overt behaviour itself. Of course the three strands, awareness, emotion and behaviour, are tightly interwoven. Language patterns that learners do not quite get correct can lead to classic errors and misconstruals of the types illustrated in section 13.2; awarenesses which are incomplete can block access to the use of an appropriate concept or technique, and so on. The three strands provide a structure which can serve, not as a mechanical 'form-filling' exercise, but as a reminder of aspects to think about when preparing to teach any topic. Hence the value in keeping notes for each topic, which can be augmented when learners display new and interesting behaviours, emotions and awarenesses, and which serve as a quick reminder of the complexity of a topic when it is time to teach it again.

The most powerful way to engage learners' full attention and involvement is to activate all three strands of their psyche, and the best way to do this is to involve learners in making choices and in using their own mathematical powers, which are discussed in the next chapter.

Reflection 13

Look back at a task you have used recently with learners, and at one of the tasks you did in an earlier chapter. In both cases, what topics might have been pertinent? What aspects of the structure of a topic emerged or do you think could have emerged if attention had been suitably directed? What other aspects of the structure of a topic would need to be brought to the surface so that learners had exposure to the full complexity of the topic?

14 Powers, Constructs and Strategies

This chapter summarises the core components of the approach to teaching and learning geometry presented in the book. Section 14.1 begins with a summary of some of the basic and natural powers which all learners possess, and which can be exploited in mathematics lessons so that learners feel that their powers are being used, rather than being suppressed or ignored. Section 14.2 summarises the pedagogic constructs which have been used explicitly and implicitly throughout the book. Section 14.3 summarises the principal pedagogic strategies used throughout the book.

14.1 MATHEMATICAL POWERS

Children demonstrate from a very young age that they have natural powers of locating relationships, proposing properties, and making and testing conjectures which involve thinking geometrically. They naturally test particular instances of generalities, and they naturally generalise (otherwise they could not use language!). They naturally make sense of their experience. Sense-making involves constructing 'stories' to account for current experiences and to integrate those with past experiences. This requires interpreting current experience in terms of past experience. Sense is made by using natural powers to collect, classify, assimilate, accommodate and reject sensations whether physical or imagined, remembered or constructed, literal or metaphoric.

Working on the tasks in previous chapters will, it is to be hoped, have reminded you of fundamental powers which you, and your learners, all possess. This section summarises and highlights these mathematical powers, drawing attention particularly to four pairs of powers which learners bring to lessons.

Imagining and Expressing

Every child can imagine, whether they use mental pictures, words, kinaesthetic responses, have a vague sort of 'sense of' or some combination of these. The power to imagine is important because it is the principal means for harnessing emotions: imagining how you wish to be and to act, in order to fill out those ideals in the future. It is by means of imagination that people are able to contemplate what is not actually present, whether it is a pattern continuing into the future or an experience from the past.

For example, imagine standing at the entrance to your kitchen; which way would you turn to get a fork? These sorts of questions can be answered because it is possible to 'be', mentally, at the entrance of your kitchen. Sometimes your body knows the answer, so you imagine yourself moving and then watch which way you moved in order to find out the answer. You may have had the experience of gazing (not staring) at a diagram and having a sense come to mind that you want to introduce a new ele-ment (complete a triangle, fill in a circle, and so on) which you realise (make real to yourself) must be present. This is how diagrams support imagery, and imagery informs the development of diagrams.

Some people use the metaphor of 'seeing' to refer to what they do in their heads, and seem to be happy that they mean 'seeing pictures in their mind's eye'. However, a significant percentage of the population does not understand what these people are talking about. They tend to work kinaesthetically or with a vague 'sense' which is very hard to describe in physical terms. Everybody, no matter what their propensities, gains from working at 'imagining' and then eventually drawing or describing what they imagine. Even when you are looking at a diagram it is important to be imagining: imagining elements that are not yet depicted, imagining what can change without changing the important relationships in the diagram, and so on.

Through mental imagery you have access to a wide variety of cases including the very large, and the very small, as in imagining the range of possible angles in a trian-gle, or the range of possible sizes of sides of a triangle. It is through the power to imagine that you access generality. Geometry is particularly powerful as a domain in which to develop the power to imagine and to express to others what you imagine. Describing what you imagine so that others can imagine or draw it is not nearly as easy as it at first seems. Learning to describe to yourself what you 'see', or imagine, what you discern, is a powerful discipline to use because it provides learners with something they can do when they get stuck in some situation.

When calling upon learners to imagine, it really helps if you cast your instructions in the imperative: imagine this, imagine that, do this, do that. After a few instructions, pause and get learners to describe what they are imagining to someone else, and when differences start to appear, perhaps a diagram can be drawn and negotiation of interpretations can begin.

The power to imagine is usefully called upon explicitly and can be developed with practice. If imagination is not called upon in mathematics, then a powerful link to the emotions is neglected, and motivation-interest may suffer. If expression in multiple forms is not encouraged, then learners may form the mistaken impression that mathe-matics does not offer opportunities for creativity. If learners encounter a very limited range of images, and a very limited range of expressions, they are likely to form the erroneous impression that mathematics is a very limited domain of human experience.

When you invite learners to close their eyes and imagine a geometrical configura-tion which you slowly develop, there are opportunities to draw attention to the freedom of choice at each stage (and the restrictions imposed by constraints, whether due to mathematical structure, or due to imposition from you). These indicate the range and scope of permitted generality, that is, possible dimensions of variation.

Specialising and Generalising

Experiencing and expressing generality is entirely natural for learners. The issue is whether that power is being called upon and developed in mathematics lessons, or whether it is being left to one side where it is likely to atrophy, at least in the mathe-matics context, through non-use. Every geometrical diagram expresses a generality, but only if you are aware of what is allowed to vary and what is not.

Many physical situations (such as door hinges, skip loaders, wine racks and sofa beds) involve geometrical relationships. A diagram helps to remove structurally irrelevant features so that relevant relationships can be depicted in a diagram. Recognising a diagram as capturing the essential structure of a physical situation is a form of generalising, as is abstracting the diagram in the first place. Learning to recognise two different diagrams as illustrations of the same relationship involves a sophisticated development from discerning different elements to recognising relationships and perceiving properties which might or might not hold when conditions vary. Something is perceived as a property only when learners become aware of what is allowed to vary, of dimensions of possible variation in the situation.

Task 14.1.1 Generalising from Quadrilaterals

Imagine four line segments which form a quadrilateral. How many different quadrilaterals can you make with those four lengths as the lengths of the sides?

Comment
After closing your eyes and imagining, and seeing that there are infinitely many, you might impose a constraint, say by specifying at least one angle, or asking what the range of possible angles can be. You could use a physical representation (straws with thread through them, hinged rods) or dynamic geometry software. You might switch from the infinity of possibilities to a more finite question such as the variety of named shapes that are possible. You might ponder what constraints four lengths have to meet in order to form a quadrilateral. You might consider what further information would be just sufficient to determine a unique quadrilateral. And so on. Each of these questions is based on some element or feature which could vary, and either constraining it or exploring its range and scope. Which well-known shapes can be converted into others by treating the joints as flexible and the angles as variable but without changing the lengths of the sides?

Now, what can be changed, what can be generalised? Can something sensible be said about five lengths forming a pentagon, or six making a hexagon? (Do not be caught by assuming a pentagon has to be regular!) This is overt generalisation of a single feature. Other things that can be varied include the relative lengths of the sides. The data could be adjusted so that instead of four side lengths you might be given three side lengths and an angle, and so on.

Seeing the general through a particular (for example, being aware of the possible variation in a diagram), and seeing the particular in the general (for example, drawing a particular diagram to illustrate a general description of a geometrical relationship such as three squares erected externally on the sides of a triangle) are core mathematical experiences. A lesson which fails to afford learners the opportunity to experience and express a generality cannot be considered to be a mathematics lesson!

Conjecturing and Convincing

Conjecturing is a way of working, an ethos, in which ideas are developed through learners thinking out loud or explicitly in some other way. *Everything* that is said is thought about and tested by those who are listening. People speak because they are uncertain and hope to get some help from others in how to articulate what they think they are 'seeing' or thinking. It is a way of working in which everyone takes responsibility for making sense of what is said by others, and anybody can be asked to explain their thinking, that is, to try to convince the rest. One of the most important things that a school can contribute in the way of developing learners' powers is to engender a conjecturing atmosphere, because of the social importance of developing a caring, listening, yet challenging way of interacting with others.

In a conjecturing atmosphere, when someone says something that is not quite understood by another, someone might ask for or offer an example; they might focus on a detail and ask pointedly for elaboration. In a conjecturing atmosphere, people do *not* say 'That's wrong', they say 'I invite you to modify your conjecture', or they say 'What about … ' and offer a possible counter-example.

Mathematical thinking really only gets going when there are competing conjectures or when there is something to justify. If learners feel that answers are always either right or wrong, they may become reticent about offering their ideas. If conjecturing is valued, and especially if modifying conjectures is valued and praised, then mathematical thinking is much more likely to flourish.

Conjecturing is about being aware of the status of some assertion: is it reasonable? Is it always true? Is it sometimes true? Is it never true? How do I know? If I cannot justify it by convincing someone else, then it remains a conjecture, something which I think may be true and for which I have some possible evidence. Being stuck on a mathematics problem is an honourable state, for it is from being stuck that challenge arises and it is from getting unstuck again that you learn how to be mathematical (using your powers to make progress). Of course, the challenge may be too great to overcome, in which case the problem can be put to one side, but only after clarifying what you know, what you are trying to prove or find, your current conjectures and any evidence you may have for them. Then it will be possible to pick up the challenge at a later date. Without this clarification you usually have to start again from scratch.

Trying to justify why you think something is the case by trying to convince someone else is most helpful in sorting out what you think, for as you start to explain it, things either tend to fall into place or fall apart. The person being explained to can learn from the experience by learning to ask probing questions ('give me an example', 'what if … was different?', 'how do you know?', 'why must … ?' and so on), which in turn develops expertise in convincing others. Thus, learning to be sceptical when listening to others trying to convince is an important part of learning to convince people yourself, for you learn to internalise the sorts of objections that others are likely to make.

When ideas are coming thick and fast it is sometimes hard to hold on to what you think is actually the case. Consequently one of the features of a conjecturing atmosphere is making a record of current conjectures. Like all mathematicians, learners sometimes run out of time when exploring some idea. A sensible place to leave off work is to make a record of current conjectures and a summary of available evidence. Developing this practice provides a satisfactory way of leaving a topic or a project and going on to something else.

Organising and Classifying

Human beings make sense by organising and classifying experience. Putting dishes and laundry away are forms of imposing order on the material world, and act as useful metaphors for working in the mental, symbolic worlds, and social worlds.

One of the reasons for organising is that it reduces confusion. It simplifies the multitude of experiences and forces which are acting upon you. The desire to impose order is manifested very early. Every experience is classified (unconsciously) in order to assimilate it into current schema and so 'make sense of it'. If it resists classification, then it is either rejected out of hand or schemas are altered in order to accommodate it (Piaget, 1971). Having acknowledged the desire and value of organising, it is useful to probe just how it is that organising comes about.

Each act of sorting involves stressing some (relevant) features and ignoring others, which in turn requires being able to discriminate those features. Sorting tasks are really excellent for getting groups of learners to express their thinking to each other and to negotiate different ways of seeing. Some learners will learn from others' ways of discerning that had not previously come to mind. Others will find their way of perceiving supported or confirmed.

The next task offers a classic context for experiencing organising and classifying, but can be seen as a generic structure for use in other contexts.

Task 14.1.2 Organising and Classifying

Write down a list of all the names for quadrilaterals that you know (for example, rhombus, or concave). Draw a diagram which indicates how the classes of quadrilateral overlap (there is overlap if there is a quadrilateral which fits both labels, such as a square being a rhombus and also a rectangle).

Draw a quadrilateral of each type for which you have a name as a description which is as general as possible yet still fits the label. Now draw in the diagonals of each and ask yourself what properties those diagonals have. Is that relationship valid for all quadrilaterals which fit the same label? Are there any others which have that property? (Adapted from Fielker, 1983.)

Comment
Discerning relationships which distinguish (what is the same and what is different) is the essence of sorting and organising. Looking for properties which all possible examples share, and which only those examples share is classifying or characterising: seeking a property which is characteristic of a class of objects. For example, parallelograms have diagonals which bisect each other, and a quadrilateral whose diagonals bisect each other must be a parallelogram. Thus, mutually bisecting diagonals characterise parallelograms from amongst all other quadrilaterals.

Did you find a class of quadrilaterals whose diagonals do not intersect? Among them are there any whose diagonals are perpendicular?

The act of sorting serves to reveal the qualities which are being used to sort, which are, by necessity, generalities, properties. Qualities or features seen as similar, relate objects in the same group together, and these arise by stressing some aspects and, consequently, ignoring others. But there is a subtle shift from seeing similarity to seeing things classified according to some property. To identify a property, and then to see if objects possess that property is quite different from being aware of relationships which, in fact, do constitute 'having that property'.

Summary

Four mathematical powers have been presented as pairs because they represent complementary experiences:

- imagining and expressing
- specialising and generalising
- conjecturing and convincing
- organising and classifying.

These are powers which can be called upon when learners are trying to make sense of mathematics. The conjecture underlying this book is that it is by prompting learners to use, and hence develop, those powers geometric thinking is best promoted.

14.2 PEDAGOGIC CONSTRUCTS

Pedagogic constructs are distinctions which have proved to be useful for thinking about teaching mathematics either

- when planning a lesson or a sequence of lessons
- in the midst of teaching, when they come to mind in the form of possible actions
- for retrospective analysis of lessons so as to learn from recent experience.

Stressing and Ignoring

Simply subjecting learners to variation may not be sufficient for them to become aware of what the teacher sees as invariant, what is permitted to change and in what way. Focusing attention on certain features places stress on them and invites consideration of whether they could actually be changed in some way, becoming a perceived dimension of possible variation. Stressing some features necessarily involves ignoring or downplaying others, and this is the essence of generalising.

Dimensions of Possible Variation (DofPV)

Any concept, whether as simple as a named shape like a rhombus, or as complicated as a theorem like Pythagoras', or a principle such as stressing and ignoring in order to provoke generalising, consists of a label and an awareness of one or more features which can be varied without altering the underlying structure or relationship. To understand and appreciate a concept is to be aware of the features of an example which can be varied and without stopping it from being an exemplar of the concept. To make sense of and use a concept it is necessary to encounter in close succession a number of variations of those aspects which can vary, and then to be in situations in which it is efficient to use the concept in order to communicate with yourself and with others. Thus the notion of *dimensions of possible variation* underpins all conceptual development.

Where there are multiple solutions (triangles with three given lines as their medians, for example) there is usually one or more dimensions of possible variation to explore: what choices are available, and what is the range of permissible change in each of those dimensions? For example, if two intersecting line segments are to be two of the medians of a triangle, then the intersection point must divide each of the segments in the ratio of 2:1 because the centroid lies two-thirds of the way from a vertex to the mid-point of the opposite side. This constraint is structural and unavoidable. Thus where the two segments intersect admits of no variation; but the angle between them, and their respective lengths are both dimensions of possible variation.

The adjective *possible* is included because different people may be aware of different possibilities. Teachers are usually aware of aspects which can be varied which may not have occurred to learners. Furthermore, the variation permitted in each dimension is referred to as the *range of permissible change*, because learners may unwittingly think that the range of change is more restricted than is actually the case.

Take, for example, the theorem that the angle subtended at the circumference of a circle by a chord is constant. When no particular care is taken to define what is meant by the 'angle subtended at the circumference', people encounter difficulties when the angle is subtended on the other side of the chord. Their range of permissible change of the point on the circumference is more limited than need be. Taking care in the definition enables the theorem to hold on the whole of the circumference: the angle subtended at a point C by a chord AB is the angle through which the line through A and C must rotate in a counter-clockwise direction about C so as to coincide with the line through B and C. Similarly, some learners may think that the sides of right angled triangles which satisfy Pythagoras' theorem have to be integers, because so much attention is devoted to that case.

The construct of *dimensions of possible variation* acts as a reminder to look for all the different features that are needed for the concept and also the features of particular examples which are not actually needed. There are close similarities to the productive questions promoted by Stephen Brown and Marian Walter (1983): 'what if ... were changed?' and 'what if not ... ?' Getting learners to reveal what they are aware of as *dimensions of possible variation* and associated *ranges of permissable change* by getting them to construct examples is a useful way to inform choices about what to work on next with learners.

Often the *dimensions of possible variation* that learners perceive and the corresponding *range of permissible change* to which they constrain their answers are not the same as those imagined by the teacher. If the teacher is talking from one view, what is said may make little sense to a learner with a different view. By being aware of this, a teacher can take action to attract learner attention appropriately .

The notion of *dimensions of possible variation* applies to tasks as well as to concepts, and indeed to the different ways in which tasks can be presented. With any task it is possible, and usually fruitful, to ask what can be changed and still the same method or approach applies. Often there are specific numbers to change, but there are other changes to be made as well. If a teacher develops and even uses explicitly the notion of *dimensions of possible variation* with respect to tasks, then learners can be encouraged to seek not just the solution to a single task, but to characterise the general class of tasks which succumb to the same technique. Learners who are aware of classes of tasks are much more likely to do well on tests than those for whom each task is new and different.

Task 14.2.1 Dimensions of Variation in Tasks

Take as the essence of a task, justifying the theorem that the angle subtended at the centre of a circle is twice the angle subtended at the circumference. What are some of the dimensions of possible variation in how the task could be presented?

Comment

The situation could be built up using mental imagery: imagine a circle; select two distinct points on the circumference and join them to the centre; join them both to a point on the circumference as well; now pay attention to the angle at the centre and the angle at the circumference: what relationship might there be?. The reasoning could also be presented as a mental imagining: join the centre to the point on the circumference; look for pairs of segments which are equal in length; look for isosceles triangles; calculate the angle at the centre in terms of the components of the angle at the circumference. Learners could be shown an interactive animation which suggests an invariance, or invited to construct the diagram themselves. Learners could be invited to draw an accurate diagram and to measure, and dynamic geometry software could be used to measure angles, though the introduction of measurement into what is a pure relationship can be misleading, for it suggests that measurement is accurate and therefore a means for justifying relationships when it is at best indicative of a possible conjecture. Learners could be given the individual statements of a reasoned justification of the theorem, but with each statement on a different card, with the task being to arrange these cards in order to make a convincing justification.

The point is that there are many different ways to present a task. When learners encounter variation in how tasks are offered they begin to be aware of a greater range of types of questions which mathematics deals with, and so they develop a broader appreciation of the nature and discipline of mathematics. If all their tasks have the same structure then boredom is likely to result, along with a restricted sense of what mathematics is about.

Doing a task is merely 'doing a task'. Prompting learners to make sense of what they have done, to consider various dimensions of possible variation and to reconstruct ideas, relationships and diagrams for themselves, all contribute to embedding their recent experience into an accessible part of memory.

Manipulating–Getting-a-Sense-of–Articulating (MGA)

As long ago as 400 BCE, Plato praised Egyptian teachers who invited learners to work with physical objects when learning arithmetic. But apparatus is only useful if it can eventually be dispensed with, otherwise learners become dependent on that particular apparatus. The point of using apparatus is to 'get a sense of' some pattern or relationship, some property or structure, and then to begin to articulate the 'sense' to yourself until it becomes fluent. The apparatus can be returned to if a situation becomes too complicated, but the apparatus is intended to provide experience and images (visual and kinaesthetic) which are internalised and integrated into learners' functioning. The notion of apparatus in geometry includes a nine-pin geoboard, a diagram (such as of seven circles as in Chapter 1), an interactive animation or use of dynamic geometry software by the learner. Drawing accurate diagrams using ruler and compass, or using dynamic geometry software, are simply auxiliary support for helping learners to 'get a sense of' some relationship, for appreciating that relationship as a potential property and, ultimately, for reasoning on the basis of that property alone.

The process of manipulating familiar, confidence-inspiring objects (which may be physical, diagrammatic or imagistic, or symbolic) in order to get a sense of relationships and invariances, in order to articulate those as properties, is fundamental to how learners actually learn. Effective tasks are ones which enable learners to get support from the teacher and from other learners in reaching succinct articulations.

Manipulating–Getting-a-Sense-of–Articulating can be seen as a spiral process of ongoing development (Figure 14.2a), in which objects which are familiar and confidence-inspiring and are manipulated in some setting in order to get a sense of some structure, enabling articulation (in diagrams, pictures, words and symbols) which becomes more and more succinct and more and more 'articulate'. These articulations (usually in symbols but also in diagrams) themselves become the components for further manipulations.

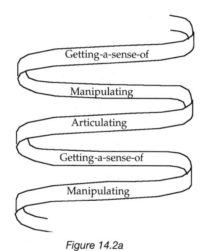

Figure 14.2a

When in the future some complexities or difficulties arise, then the learner has recourse to confidently manipulable objects through which to regain a sense of structure and so regain confidence. Whenever you encounter a statement that is not very clear, perhaps because it is very general or because it uses technical terms, it can help to turn to some example which is confidently manipulable. This may take the form of something physical, or it may consist of symbols, as long as you are *confident in using those symbols*. Comments and suggestions on some of the tasks have tried to encourage you to do this, but also to be aware of just how powerful and effective it can be as a strategy.

Do–Talk–Record (DTR)

Manipulating is not learning, merely opportunity to learn and preparation for learning. Doing tasks is not learning, only an opportunity for learning. Getting learners to talk about what they are doing, trying to articulate to others the sense they are making, contributes to the possibility of making some sort of record of what they are doing. That record may use diagrams, pictures, words and symbols. Trying to record can inform your talk; talking can inform what and how you 'do'. What is important, and what is signalled by the DTR triple construct, is the importance of paying attention to how you do something, not just to getting it done.

Discussion between learners can be highly effective, as long as it is mathematical in nature, that is, as long as it is supported by a conjecturing atmosphere in which everyone listens, constructs examples and possible counter-examples, and tries to articulate for themselves. Only when it becomes convenient to use labels to refer to objects, whether present or only imagined, known or unknown, is it necessary and appropriate to move to formal labels (symbols).

The triple construct Doing–Talking–Recording can act as a reminder that the point of doing is to get better at articulating something, that recording emerges out of articulation as well as feeding ever improved articulation, and that articulation emerges out of paying to attention to how you are doing something, as well as improving that doing through articulating.

Enactive–Iconic–Symbolic and Different Worlds

Jerome Bruner developed the terms *enactive*, *iconic* and *symbolic* to refer to different forms of representation. Each is like its own world, involving ways of thinking. A fourth world, which Bruner drew attention to later in his career, is the world of social practices and interpersonal relations. People operate within this fourth world through representations drawn from the other three. Thus people demonstrate with objects, describe and draw images, and use formal symbols. The social is where people encounter and become enculturated into the use of objects, diagrams and descriptions, formal symbols, and ways of working and communicating with themselves and with each other.

Rather than preferring one world over others, or letting one world dominate others, what is most valuable is achieving a balance among all four. Practices are picked up through being in the presence of other people using them. For example, in classrooms in which 'same and different' are used frequently and effectively, learners pick them up; in classes where conjecturing is the norm, learners pick up the associated practices; in classes where individual methods are valued but also challenged and stretched in order to be more efficient, learners actually become more mathematically sophisticated and mature.

See–Experience–Master (SEM)

When a new idea is encountered, it is not always taken in immediately. For example, when you encounter a new word, you may use it in too broad and general a meaning, before it contracts back into common usage. So, too, meeting new ideas is sometimes like experiencing a train rushing through a station: lots of excitement and noise, but not a lot of detail grasped. With continued exposure over time, experience becomes richer and more detailed. After a while, that experience includes increasing competence and fluency, leading to 'mastery' of the techniques and familiarity with the ideas. 'Doing things', manipulating familiar and confidence-inspiring objects, getting a sense of structural patterns, relationships and properties, and articulating these with increasing succinctness and fluency is how substantial learning comes about.

By giving each of the three blocks of this book a parallel structure, often returning to the same theme but at a more sophisticated level, opportunities were offered not only to see ideas go by, but to re-encounter them several times in different ways. This mirrors how people learn naturally in situations outside school as well.

Structure of Attention

When confronted with something unfamiliar, attention is either on the whole or on some part of that whole. After what may be a moment or a long time, details start to emerge: some aspects or parts are distinguished or discerned from among the rest. Put another way, at first, there are details which are either not seen at all or ignored, as when learners are invited to say what they see in a diagram but do not make any reference to shapes beyond triangles and squares.

Task 14.2.2 Attention!

Say (to yourself, to a friend, to a goldfish!) what you see that is the same, and what different in and between the three diagrams in Figure 14.2b. Start simply!

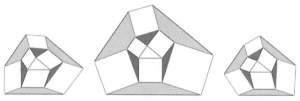

Figure 14.2b

What do you notice about how your attention moves, about what you attend to?

Comment
Did you notice that the central triangle is the same? Did you detect relationships such as parallel segments? Did you detect possible squares in the middle diagram? Did comparing the three reinforce and support the notion that parallelism seems to be involved? Did you become aware of sensing one or more dimensions of possible variation? Were you aware of checking discerned sub-diagrams for properties (such as being a square, parallelogram or a trapezium)? Did you think in terms of the Vecten diagram in the case of the middle one? Did that lead you to see the others as variations or extensions?

Until you discern details, which involves focusing attention, you cannot do anything but be aware of the whole. As you focus on a detail (perhaps one of the diagrams or some part of one of them), discerning it from what is around it (stressing and consequently ignoring) you become aware of relationships. Here, there are relationships within a part such as parallelism and equality of length or angle, and relationships between the three diagrams, such as a central triangle surrounded by parallelograms (in a special case, squares), with flanking triangles and then in the next layer, shapes that seem to be trapeziums. The Vecten-like diagram in the middle might prompt you to conjecture that the three shaded triangles are equal in area, even if they are not equal to the central triangle in all three cases.

Your attention naturally shifts between seeking relationships and becoming aware of properties that can be checked: what is needed to be sure a shape is a parallelogram? A trapezium? Attention to properties may enable you to reason on the basis of properties: in the Vecten diagram there are several ways of seeing the equality of area (trigonometry, rotating pieces of triangle), so perhaps the same reasoning can be used more generally.

Often when gazing at a diagram (not just scanning for details, but trying to get a sense of the whole), you find yourself discerning an element which is not actually present: perhaps four points are cyclic but the circle is missing; perhaps filling in the triangle made from three points might reveal some useful angles; or a segment with its mid-point would help you to check out a possible relationship. Thus discernment is not just of what is, but of what could be, and is intimately tied up with seeking relationships and perceiving or applying properties.

The upshot of all this is that it illustrates how human attention shifts rapidly between different ways of attending and different foci of attention. Sometimes there are several of these going on either simultaneously, or in swift succession. In summary, there can be:

- awareness of or focus on wholeness: the identification of an object (which may be a part of some other object);
- a shift to awareness of or focus on discerned details comprising that object (creating sub-objects);
- a shift to awareness of relationships or similarities between features comprising sub-objects at any level of detail;
- a shift to awareness of or focus on properties that (sub-)objects might satisfy;
- a shift to awareness of or focus on properties as the sole basis from which deductions can be made.

These different forms of attention are similar to a series of five levels of geometric thinking proposed by Dina van Hiele-Geldof and Pierre van Hiele (1986; see also Burger and Shaughnessy, 1986; Usiskin, 1982): visualisation, analysis, abstraction, deduction and rigour. However the suggestion being made in this book is that shifts between these rather subtle differences in attention are going on all the time and not necessarily in any specific sequence. Working on a task, everyone shifts between the van Hiele states, often very quickly. Sometimes in a given situation learners get stuck, and while others are talking from one point of view, say about relationships, or about properties independent of the particular, the learners are still discerning details.

Attention is too will o' the wisp to be subject to predetermined sequences. By being aware of these shifts within yourself, you are better attuned to recognising shifts that learners are, or perhaps are not yet, making. Such awareness enables you to be of more specific assistance to them through directing their attention or being patient until they are in a position to shift their attention than if you are oblivious to the subtle shifts in the structure of their attention.

What Makes an Example Exemplary?

When presenting an example to learners, whether illustrating a concept or property, or showing how to use a technique in a particular case, it is worthwhile paying attention to which features are generic and to be treated as a dimension of possble variation, which aspects of the range of permissable change are indicated, and which features are extraneous and irrelevant. Emphasis can then be placed on the important ones through the use of colour and voice tones. Offering several examples can be more helpful, but only if learners appreciate what is the same as well as what is different about the various examples. In other words, are the various dimensions of possible variation being highlighted by not varying too many features at once, and are the ranges of permissable change sufficiently clearly exposed?

A particular square is only an example of a square, or a rhombus, or a parallelogram, or a trapezium or a quadrilateral (or even a pentagon with one side of length zero!) if you 'see it' that way, through stressing some aspects and ignoring others, through being aware of possible variation.

Didactic Transposition

Yves Chevellard (1985, see Mason and Johnston-Wilder, 2004) drew attention to a very common phenomenon, which is almost an endemic tension. An expert is stimulated by discovering some relationship or connection, some technique or way of thinking. So they construct some tasks which are intended to enable learners to have a similar experience. The trouble is that the expert's awareness has been transformed or transposed into instruction in behaviour: do this, do that. The tasks in this book all suffer from this transposition to a greater or lesser extent. That is why the tasks themselves are of limited value. What makes any of them effective is the reflection, the reconstruction, the re-articulation which you have gone through for yourself. It is only when you begin to explain something to someone else that you really sort out how to articulate your awareness succinctly and effectively. Assuming that the listener then appreciates what you were experiencing is one of the fundamental flaws in education. Learners usually experience neither what the teacher experienced nor what the mathematical pioneers experienced when the ideas were first being developed. But by struggling to reconstruct and to explain to others, learners can build on the experience of undertaking a task in order to reach some measure of understanding. Being asked to construct examples, perhaps meeting specified constraints, then reveals the extent of that awareness, the dimensions of possible variation and associated ranges of permissable change of which the learner is currently aware.

14.3 PEDAGOGIC STRATEGIES

This section reiterates some of the pedagogic strategies suggested by the constructs in the previous section, and used or suggested throughout the book.

Say What You See

Presenting learners with a diagram, a set of exercises or a worked example, and asking them simply to 'say what you see' (without using technical terms or trying to be clever) can be very instructive. It helps learners to take in the whole, to discern details and to learn from details others have discerned. Where someone describes something that others are unclear about, there is opportunity to reinforce a conjecturing atmosphere and to get learners to work on asking specific and pointed questions, instead of 'say that again'. Asking each person to say just one thing each in turn, enables many or most to contribute something. Doing all this without physically pointing can strengthen learners' control over both their mental imagery and their verbal descriptions of what is present in their attention. Over time, 'say what you see' can become something that individuals do for and with themselves as they tackle the unfamiliar and the complex.

When learners are shown an animation or a mathematical film, they can be asked to reconstruct the sequence of movements collectively, each contributing something of what they saw. It is usual that some people report seeing things that others do not recall seeing: then a re-showing has a purpose, as people check out details they missed the first time.

When trying to detect underlying structure or pattern it often helps to try some particular cases for yourself. This is an example of specialising, and as with any specialising, what matters is not the answers, but attending to what you actually do. This means monitoring your actions, especially the ones that spring from nowhere, as it were. It does not mean slowing down and being careful but, rather, maintaining speed while being awake to what you are doing. Thus an instruction to 'copy and complete' a table is most likely to turn into a clerical exercise in which learners pay little or no attention to what is going on. Doing one or two special cases with full attention to the 'how', perhaps even tracking the arithmetic rather than doing it all, is often much more informative than a page full of calculations all done mechanically.

Same and Different

A useful question to initiate mathematical thinking is to ask learners what they see as the same, and what different, about two or more objects (as in Task 14.2.2). These objects could be numbers, problems, diagrams, physical objects, sequences of numbers or objects, and so on. For example, asking what is the same and what different about two particular triangles can lead to seeing that they are congruent or similar; about two angles can lead to seeing that they are both subtended from the same chord of a potential or actual circle, and so on. The aim of the question is to draw attention to aspects that are different, that can change and, at the same time, to aspects which are invariant and so do not change, suggesting a structural relationship. This supports awareness and articulation of generality.

Asking learners 'what is the same and what different about' (say, two or more objects or two or more figures or sub-figures) is often fruitful for it reveals some of what they are discerning and the relationships they are recognising. It passes initiative to learners; it exercises their attention and their control of that attention, including the power of mental imagery; it provides the basis for the exercise of powers of specialising and generalising; and it reinforces the mathematical theme of invariance in the midst of change.

Laurinda Brown and Alf Coles (2000) demonstrated that when used frequently, looking for similarities and differences ('same 'n' different') can be internalised and used spontaneously by learners.

Another and Another

A good way to encourage creativity and playfulness is to ask learners to construct an object meeting certain constraints, then another and then another, knowing that they are not going to be asked to do anything more complicated with whatever they construct. Some will be carefully conservative in case they are then asked to do something tricky with their example, but after a while they will see that this is not to be the case. When they hear other people's ideas some will resolve to be more adventurous next time. Asking for more extreme examples which illustrate a relationship or theorem, or asking for what relationships or properties might be being illustrated in a specific diagram offers learners the opportunity to discern, relate and perceive properties. Asking for objects which meet increasingly restrictive constraints enables learners to appreciate the impact on freedom of choice that additional constraints are likely to have, thereby supporting their sense of what a variable is.

In How Many Ways?

Some of the tasks in this book asked 'in how many ways can you … ?' rather than the more usual 'can you find … ?' or simply 'find a … ' task structure which is so familiar in textbooks. Studies of differences between Japanese classrooms and English-speaking classrooms revealed a prevalence of 'in how many ways?' in Japanese classrooms rather than the more direct 'find' that tended to feature in English-speaking classrooms (Stigler and Hiebert, 1999). The more open question invites looking for multiple approaches and reinforces creative and constructive aspects of geometrical thinking, whereas the search for the 'correct answer' is so much more limiting, and gives an inappropriate picture of what mathematics is actually like.

Turn a Doing into an Undoing

Most tasks, no matter how routine, can be transformed into challenging tasks requiring insight and creativity simply by converting a 'doing' into an 'undoing'. Since every triangle has three medians, three altitudes, three angle bisectors (not including exterior angle bisectors), three right bisectors, to name just a few, there are numerous opportunities for trying to reconstruct a triangle given the three medians as lines, or as segment lengths, the three altitudes as lines, or as segment lengths, and so on. Often this leads to multiple solutions and the need for a construction or creation which is absent in the reverse direction. For example, the mid-points of the edges of a quadrilateral form a parallelogram: find all the quadrilaterals which have a given parallelogram as the parallelogram of mid-points; given a triangle, for what triangles is it the triangle of mid-points of the sides (or of feet of altitudes, or …)?

Scaffolding and Fading

Jerome Bruner (1986) and colleagues drew attention to the fact that learners' attention is often fully taken up with details of a computation or a problem, and that the role of the teacher is to be 'consciousness for two', holding onto awareness of the larger goal and not getting lost in the details. So arose the notion of *scaffolding* to refer to the support provided by the teacher. For example, if a teacher asks the same question repeatedly over several weeks, such as 'what is the same and what different?', 'what is changing and what is staying the same?' or, when learners are stuck, 'have you tried a simpler example?', learners may find themselves prompted to make progress.

However, learners may easily become dependent on teacher prompts, and may not even be aware of the form or nature of those prompts. It is necessary therefore to make the prompts gradually less and less direct. For example, shifting from 'can you give me an example?' or 'what might be a particular case?', to, 'what did I suggest last time?', 'what question do you think I am going to ask you?' or 'what did we do last time this happened?' prompts the learner to become aware of the teacher's scaffolding, and thus to internalise it for their own use. At first such 'meta-prompts' may result in learners being somewhat taken aback and bemused, but they soon work out what is being asked of them. Eventually the prompts become so indirect and so infrequent that learners find themselves using them spontaneously. Many authors independently pointed out that what is important about scaffolding as prompts to support learners is the fading of the prompts, so that learners begin to use them spontaneously for themselves.

Learner-Constructed Examples

Learners can be given opportunities to express their creativity and to make choices for themselves by being asked to construct objects meeting certain constraints. In the process, learners display some of the dimensions of possible variation and associated ranges of permissible change of which they are aware. Asking them to construct another and another (see above) may prompt them to explore the boundaries of their confidence and so extend the range of examples of which they are aware and with which they are confident (Watson and Mason, 2002; 2005).

When learners move from being satisfied merely to complete an assignment, to being confident they could solve a similar problem in the future, they are beginning to educate their awareness concerning the topic as a whole, and not just training their behaviour in the mechanics of techniques. When learners can construct sample tasks for themselves which illustrate their prowess in solving a class of problems, they are making global sense rather than simply getting through the work. When learners can describe or even express a general class of geometrical objects which have a specified property, they are enabling themselves to reconstruct the property and its consequences on some future occasion. They are more likely to perform well on tests and examinations, even when they meet unfamiliar problems.

Diverting Attention in Order to Automate

An expert is someone who does not need to place their full attention in the carrying out of that expertise. Caleb Gattegno (1987) argued that in order to develop competence and fluency it is necessary to divert your conscious attention away from what you are doing, rather than into it. Dave Hewitt (1996) developed this idea, suggesting that by setting learners tasks which involve them in specialising by constructing their own examples which require the use of a technique to be automated or a relationship or property, learner attention can be drawn away from the doing in order to keep track of what the results of the doing say about the task in hand. Moving from novice to expert means integrating the technique so that performing it requires less and less attention and enabling more attention to be directed towards larger goals.

This book does not offer practice exercises; instead you have been invited to engage in ever more sophisticated tasks and challenges, with a view to developing your competence and fluency through use while meeting further challenges, rather than in repetition of tasks of the type already encountered. It has been assumed that where you have felt the need for more routine practice, you can construct it for yourself, but that competence and fluency develop through meaningful use in context not from mindless rehearsal.

Teaching Techniques

It is very tempting to isolate a technique which will serve learners well on examinations, and then teach them the technique through worked examples and plenty of practice. But in an examination it is necessary for learners first to recognise each question as belonging to a type, and then to have the appropriate technique come to mind. In geometry, they need to discern relevant parts of a diagram, seek out relevant relationships, make use of properties and reason on the basis of those properties. If learners are led, through suitably constructed tasks, to construct viable and efficient methods for themselves (usually through discussion and reflection) then they are more likely to remember the technique or to be able to reconstruct it when needed, and more likely to recognise its relevance. They are also more likely to be able to adapt a technique to a novel situation than if they have been trained in the specifics of a technique.

Reflection 14

Think back to a task you have worked on recently. What mathematical powers were you aware of using? What other powers might you have used to help?

What could you do to get learners to make more use of their own powers?

Epilogue

You have been invited throughout the book to integrate encounters with some of the important aspects of geometry with ways of interacting with learners. The overall design has been to engage you in tasks in order to accumulate experience which can then be used to make sense of pedagogic remarks. Only after thoughtful reflection does it make sense to think about implications for teaching. Through engaging in geometric thinking, with frequent invitations to pause and reflect, you have been prompted to reconsider and reconstruct what has been experienced. This approach to professional development is summarised by the triple Adult–Process–Classroom (APC for short), and is recommended as a structure for working with colleagues, as well as when working with learners in classrooms. Start from recent experience; use that experience to make sense of past experience by reflecting on processes; re-enter significant and salient moments; recall details and, most especially, imagine yourself in a typical classroom situation making use of a task type or strategy. In the case of learners, re-entering significant moments and reconstructing portions of reasoning, diagram construction, relationships and properties, and imagining themselves encountering a situation in which the same ideas or techniques are pertinent by discerning the DofPV and RofPCh of the tasks undertaken. Working in this way, striving to be more disciplined in noting actions to take and things to look out for in classrooms, and in planning for the future by imagining yourself acting the way you would like to, is the beginning of a science of education. Caleb Gattegno (1970; 1987) and Hans Freudenthal (1983) both strove to outline such a science, but their writings can at best be inspiration. The real science is what you choose to do, yourself, in your situation.

I cannot change others;

I can however work at changing myself.   (Mason, 2002, Frontispiece)

It is amazing what an influence this can have on others.

References

Askew, M., Brown, M., Rhodes, V., Johnson, D. and William, D. (1997) *Effective Teachers of Numeracy*. London: Kings College.

ATM mats (webref) available from ATM Derby, http://www.atm.org.uk (accessed April 2005).

Ausubel, D.P. (1963) *The Psychology of Meaningful Verbal Learning: An Introduction to School Learning*. New York: Grune and Stratton.

Bates, T. (1979) 'Triangles and squares', *Mathematics Teaching*, 88: 38–41.

Bell, E. (1953) *Men of Mathematics*. London: Penguin.

Bishop, A. (1991) *Mathematical Enculturation*. Dordrecht: Kluwer Academic.

Black, M. (1993) 'More about metaphor', in A. Orton (ed.), *Metaphor and Thought*. 2nd edn. Cambridge: Cambridge University Press.

Brown, L. and Coles, A. (2000) 'Same/different: a "natural" way of learning mathematics', in T. Nakahara and M. Koyama (eds), *Proceedings of the 24th Conference of the International Group for the Psychology of Mathematics Education*, Vol. 2. Hiroshima: Hiroshima University, pp. 153-60.

Brown, L., Hewitt, D. and Tahta, D. (eds) (undated) *A Gattegno Anthology: selected articles by Caleb Gattegno reprinted from Mathematics Teaching*. Derby: Association of Teachers of Mathematics.

Brown, S. and Walter, M. (1983) *The Art of Problem Posing*. Philadelphia, PA: Franklin Press.

Bruner, J. (1986) *Actual Minds, Possible Worlds*. Cambridge, MA: Harvard University Press.

Burger, W.F. and Shaughnessy, J.M. (1986) 'Characterising the van Hiele levels of development in geometry', *Journal for Research in Mathematics Education*, 17(1): 31–48.

Chevallard, Y. (1985) *La Transposition Didactique*. Grenoble: La Pensée Sauvage.

Cundy, H. and Rollett, A. (1981) *Mathematical Models*. 3rd edn. Stradbroke: Tarquin.

Cuoco, A., Goldenberg, E.P. and Mark, J. (webref) 'Geometric approaches to things' in *Habits of Mind: an organising principle for mathematics curriculum* available at http://www.edc.org/MLT/ConnGeo/HOM.html

Cut-the-Knot (webref) http:www.cut-the-knot.org (accessed April 2005).

De Villiers, M. (1990) 'The role and function of proof in mathematics', *Pythagoras*, 24: 17–24.

Ellis, H.F. (1981) *A.J. Wentworth, B.A.* London: Arrow Books.

Fielker, D. (1983) *Removing the Shackles of Euclid*. Derby: Association of Teachers of Mathematics.

Franco, B. (1999) *Unfolding Mathematics with Unit Origami*. Berkeley, CA: Key Curriculum Press.

Freudenthal, H. (1983) *Didactical Phenomenology of Mathematical Structures*. Dordrecht: Reidel.

Freudenthal, H. (1991) *Revisiting Mathematics Education: China Lectures*. Dordrecht: Kluwer Academic.

Gattegno, C. (1970) *What We Owe Children: The Subordination of Teaching to Learning*. London: Routledge and Kegan Paul.

Gattegno, C. (1987) *The Science of Education Part I: Theoretical Considerations*. New York: Educational Solutions.

Gaudi (webref) Casa Milà http://www.LAPEDRERAeducacio.org (accessed April 2005).

Gerdus, P. (1988) 'On culture, geometric thinking and mathematical education', *Educational Studies in Mathematics*, 19: 137–62.

Hardy, G.H. (1941) *A Mathematician's Apology*. Cambridge: Cambridge University Press. 2005 edition available as pdf at www.math.ualberta.ca/~mss/books/A Mathematician's Apology.pdf (accessed June 2005).

Hewitt, D. (1996) 'Mathematical fluency: the nature of practice and the role of subordination', *For the Learning of Mathematics*, 16(2): 28–35.

Hoyles, C. and Healy, L. (1997) 'Unfolding meanings for reflective symmetry', *International Journal of Computers in Mathematical Learning*, 2: 27–59.

Hoyles, C. and Jones, K. (1998) 'Proof in dynamic geometry contexts', in C. Mammana and V.Villani (eds), *Perspectives on the Teaching of Geometry for the 21st Century*. Dordrecht: Kluwer Academic. pp. 121–28.

Hoyles, C., Küchemann, D. and Foxman, D. (2003a) 'Comparing geometry curricula: insights for policy and practice', *Mathematics in School*, 32 (3): 2–6.

Hoyles, C., Küchemann, D. and Foxman, D. (2003b) 'The role of proof in different geometry curricula'. *Mathematics in School*, 32 (4): 36–40.

Kanizsa figures (webref) http://mathworld.wolfram.com/KanizsaTriangle.html (accessed April 2005).

King, J.R. (1996) *Geometry Through the Circle with the Geometer's Sketchpad*. Berkeley, CA: Key Curriculum Press.

Lenart, I. (1996) *Non-Euclidean Adventures on the Lenart sphere*. Emeryville, CA: Key Curriculum Press.

Lopez-Real, F. (2003) 'Collaborative reflection through sharing significant incidents', in A. Peter-Koop, V. Santos-Wagner, C. Breen and A. Begg (eds), *Collaboration in Teacher Education: Examples from the Context of Mathematics Education*. Dordrecht: Kluwer Academic. pp. 235–51.

Maor, E. (1998) *Trigonometric Delights*. Cambridge, MA: Princeton University Press. Available in pdf format at http://www.pupress.princeton.edu/books/Maor (accessed April 2005).

Mason, J. (1989) 'Geometry: what, why, where and how?', *Mathematics Teaching*, 129: 40–47.

Mason, J. (2002) *Researching Your Own Practice: The Discipline of Noticing*. London: RoutledgeFalmer.

Mason, J. and Johnston-Wilder, S. (2004) *Fundamental Constructs in Mathematics Education*. London: RoutledgeFalmer.

Mathworld (webref) http://mathworld.wolfram.com/SphericalGeometry.html (accessed April 2005).

McLeay, H. (2002) 'Geostrips or dynamic geometry', *Micromath*, 18(3): 7–10.

NRICH (webref) http://nrich.maths.org.uk/public/(accessed April 2005).

Piaget, J. (1971) *Biology and Knowledge: An Essay on the Relations between Organic Regulations and Cognitive Processes*. Chicago, IL: University of Chicago Press (originally published in French, 1963).

Pirie, S and Kieren, T. (1994) 'Growth in mathematical understanding: how can we characterise it and how can we represent it?', *Educational Studies in Mathematics*, 26(2–3): 165–90.

Polya, G. (1957) *How to Solve It: A New Aspect of Mathematical Method*. (2nd edn.). Cambridge, MA: Princeton University Press.

Pythagoras (webref) http://www.sunsite.ubc.ca/LivingMathematics/V001N01/UBCExamples/Pythagoras/pythagoras.html (accessed April 2005)

Riley, B. (1961) *Movement in Squares*. Tempera on Hardboard. More information available at http://www.mishabittleston.com/artists/bridget-riley (accessed April 2005).

Senechal, M. (1990) 'Shape' in *On the Shoulders of Giants: New Approaches to Numeracy*, Lynn Athur Steen (ed.) Washington, D.C: National Academy Press. pp. 139–81

St Andrews (webref) 'The MacTutor History of Mathematics archive', http://www-groups.dcs.st-and.ac.uk/~history (accessed April 2005).

Stigler, J. and Hiebert, J. (1999) *The Teaching Gap: Best Ideas from the World's Teachers for Improving Education in the Classroom*. New York: Free Press.

Swetz, F. and Kao, T. (1977) *Was Pythagoras Chinese? An examination of Right Triangle Theory in Ancient China*. Harrisburg, PA: Pennsylvania State University Press.

Tangram puzzles (webref) http://www.myweb3000.com/tangramgame.html (accessed April 2005).

Usiskin, Z. (1982) *Van Hiele Levels and Achievement in Secondary School Geometry*. Chicago, IL: University of Chicago Press.

Van Hiele, P. and van-Hiele-Geldof, D. (1986) *Structure and Insight: A Theory of Mathematics Education*. Developmental Psychology Series. London: Academic Press.

Watson, A. and Mason, J. (2002) 'Student-generated examples in the learning of mathematics', *Canadian Journal of Science, Mathematics and Technology Education*, 2(2): 237–49.

Watson, A. and Mason, J. (2005) *Mathematics as a Constructive Activity: The Role of Learner-Generated Examples*. Mahwah, NJ: Erlbaum.

Wills, H. (1985) *Leonardo's Dessert, No Pi*. Reston, VA: NCTM.

Index